Programming 16-Bit PIC Microcontrollers in C

Programming 16-Bit PIC Microcontrollers in C

Learning to Fly the PIC24

By

Lucio Di Jasio

AMSTERDAM • BOSTON • HEIDELBERG • LONDON
NEW YORK • OXFORD • PARIS • SAN DIEGO
SAN FRANCISCO • SINGAPORE • SYDNEY • TOKYO

Newnes is an imprint of Elsevier

ELSEVIER

Newnes

Newnes is an imprint of Elsevier
30 Corporate Drive, Suite 400, Burlington, MA 01803, USA
Linacre House, Jordan Hill, Oxford OX2 8DP, UK

 Recognizing the importance of preserving what has been written,
Elsevier prints its books on acid-free paper whenever possible.

Library of Congress Cataloging-in-Publication Data

(Application submitted.)

British Library Cataloguing-in-Publication Data
A catalogue record for this book is available from the British Library.

ISBN-13: 978-0-7506-8292-3
ISBN-10: 0-7506-8292-2

For information on all Newnes publications
visit our Web site at www.books.elsevier.com

07 08 09 10   10 9 8 7 6 5 4 3 2 1

Printed in the United States of America

Dedication

To Sara

Contents

Contents

Contents

Contents

Preface

Writing this book turned out to be much more work than I had expected and, believe me, I was already expecting a lot. This project would never have been possible if I did not have 110% support and understanding from my wife, Sara. Special thanks also go to Steve Bowling, a friend, a pilot and an expert on Microchip 16-bit architecture, for reviewing the technical content of this book and providing many helpful suggestions for the demonstration projects and hardware experiments. Many thanks go to Eric Lawson for constantly encouraging me to write and for all the time he spent fixing my eternally long-running sentences and my bad use of punctuation. I owe big thanks also to Thang Nguyen, who was first to launch the idea of the book; Joe Drzewiecky and Vince Sheard for patiently listening to my frequent laments and, always working hard on making MPLAB® IDE a better tool; Calum Wilkie and Guy McCarthy for quickly addressing all my questions and offering so much help and insight into the inner workings of the MPLAB C30 compiler and libraries. I would also like to extend my gratitude to all my friends and colleagues at Microchip Technology and the many embedded-control engineers I have been honored to work with over the years. You have so profoundly influenced my work and shaped my experience in the fantastic world of embedded control.

Introduction

The story goes that I badly wanted to write a book about one of the greatest passions in my life: flying! I wanted to write a book that would convince other engineers like me to take the challenge and live the dream—learn to fly and become private pilots. However, I knew the few hours of actual flying experience I had did not qualify me as a credible expert on the art of flying. So when I had an opportunity to write a book about Microchip's new 16-bit PIC24 microcontrollers, I just could not resist the temptation to join the two things, programming and flying, in one project. After all, learning to fly means following a well-structured process—a journey that allows you to acquire new capabilities and push beyond your limits. It gradually takes you through a number of both theoretical and practical subjects, and culminates with the delivery of the private pilot license. The pilot license, though, is really just the beginning of a whole new adventure—a license to learn, as they say. This compares very well to the process of learning new programming skills, or learning to take advantage of the capabilities of a new microcontroller architecture.

Throughout the book, I will make brief parallels between the two worlds and in the references for each chapter I will add, here and there, some suggestions for reading about flying. I hope I will stimulate your curiosity and, if you happen to have this dream inside you, I will give you that last final push to help make it happen.

Who should read this book?

This is the part where I am supposed to tell you that you will have a wonderful experience reading this book, that you will have a lot of fun experimenting with the software and hardware projects, and, that you will learn about programming a shiny new 16-bit RISC processor in C, practically from scratch. But, in all honesty, I cannot! This is only partially true. I do hope you will have a lot of fun reading this book and the experiments are…"playful," and you should enjoy them. However, you will need some preparation and hard work in order to be able to digest the material I am presenting at a pace that will accelerate rapidly through the first few chapters.

This book is meant for programmers having a basic to intermediate level of experience, but not for "absolute" beginners. Don't expect me to start with the basics of binary numbers, hexadecimal notation or the fundamentals of programming. However, we will briefly review the basics of C programming as it relates to applications for the latest generation of general-purpose 16-bit microcontrollers, before moving on to more challenging projects. My assumption is that you, the reader, belong to one of four categories:

- Embedded-control programmer: experienced in assembly-language microcontroller programming, but with only a basic understanding of the C language.

- PIC® microcontroller expert: having a basic understanding of the C language.

- Student or professional: with some knowledge of C (or C++) programming for PCs.

- Other SLF (superior life forms): I know programmers don't like to be classified that easily, so I created this special category just for you!

Depending on your level and type of experience, you should be able to find something of interest in every chapter. I worked hard to make sure that every one of them contained both C programming techniques and new hardware-peripheral details. Should you already be familiar with both, feel free to skip to the experts section at the end of the chapter, or consider the additional exercises, book references and links for further research/reading.

These are some of the things you will learn:

- The structure of an embedded-control C program: loops, loops and more loops

- Basic timing and I/O operations

- Basic embedded-control multitasking in C, using the PIC24 interrupts

- New PIC24 peripherals, in no specific order:
  - Input Capture
  - Output Compare
  - Change Notification
  - Parallel Master Port
  - Asynchronous Serial Communication
  - Synchronous Serial Communication
  - Analog-to-Digital Conversion

- How to control LCD displays

- How to generate video signals

- How to generate audio signals

- How to access mass-storage media

- How to share files on a mass-storage device with a PC

Structure of the book

Similar to a flying course, the book is composed of three parts. The first part contains five small chapters of increasing levels of complexity. In each chapter, we will review one basic hardware peripheral of the PIC24FJ128GA010 microcontroller and one aspect of the C language, using the MPLAB C30 compiler (Student Version included on the CD-ROM). In each chapter, we will develop at least one demonstration project. Initially, such projects will require exclusive use of the MPLAB SIM software simulator (included on the CD-ROM), and no actual hardware will be necessary, although an Explorer 16 demonstration board might be used.

In the second part of the book, containing five more chapters, an Explorer16 demonstration board (or third-party equivalent) will become more critical, as some of the peripherals used will require real hardware to be properly tested.

The third part of the book contains five larger chapters. Each one of them builds on the lessons learned in multiple previous chapters, while adding new peripherals to develop projects of greater complexity. The projects in the third part of the book require the use of the Explorer 16 demonstration board and basic prototyping-skills, too (yes, you might need to use a soldering iron). If you don't want to or you don't have access to basic hardware-prototyping tools, an ad hoc expansion board containing all the circuitry and components necessary to complete all the demonstration projects will be made available on the companion Web site: *http://www.flyingthepic24.com*.

All the source code developed in each chapter is also available for immediate use on the companion CD-ROM.

What this book is not

This book is not a replacement for the PIC24 datasheet, reference manual and programmer's manual published by Microchip Technology. It is also not a replacement for the MPLAB C30 compiler user's guide, and all the libraries and related software tools offered by Microchip. Copies are available on the companion CD-ROM, but I expect you to download the most recent versions of all those documents and tools from Microchip's Web site (*http://www.microchip.com*). Familiarize yourself with them and keep them handy. I will often refer to them throughout the book, and I might present small block diagrams and other excerpts here and there as necessary. However, my narration cannot replace the information presented in the official manuals. Should you notice a conflict between my narration and the official documentation, ALWAYS refer to the latter. Please do send me an email if a conflict arises. I will appreciate your help and I will publish any corrections and useful hints I receive on the companion Web site: *http://www.flyingthepic24.com*.

This book is also not a primer on the C language. Although a review of the language is given throughout the first few chapters, the reader will find in the references several suggestions on more complete introductory courses and books on the subject.

Checklists

Pilots, both professional and not, use checklists to perform every single procedure before and during a flight. This is not because the procedures are too long to be memorized or because pilots suffer from more memory problems than others. They use checklists because it is proven that the human memory can fail and that it tends to do so more often when stress is involved. Pilots can perhaps afford fewer mistakes than other proffessionals, and they value safety above their pride.

There is nothing really dangerous that you as a programmer can do or forget to do while developing code for the PIC24. Nonetheless, I have prepared a number of simple checklists to help you perform the most common programming and debugging tasks. Hopefully, they will help you in the early stages, when learning to use the new PIC24 toolset—even later if you are, like most of us, alternating between several projects and development environments from different vendors.

PART

I

The first flight

In This Chapter

- ▶ *Compiling and linking*
- ▶ *Building the first project*
- ▶ *PORT initialization*
- ▶ *Retesting PORTA*
- ▶ *Testing PORTB*

The first flight for every student pilot is typically a blur—a sequence of brief but very intense sensations, including:

- The rush of the first take-off, which is performed by the instructor.

- The white-knuckled, sweaty grip on the yoke while trying to keep the plane flying straight for a couple of minutes, after the instructor gives the standard "anybody that can drive a car can do this" speech.

- Acute motion sickness, as the instructor returns for the landing and performs a sickness-inducing maneuver, called the "side slip." where it looks like the runway is coming through the side window.

For those who are new to the world of embedded programming, this first chapter will be no different.

Flight plan

Every flight should have a purpose, and preparing a flight plan is the best way to start.

This is going to be our first project with the PIC24 16-bit microcontroller and, for some of you, the first project with the MPLAB® IDE Integrated Development Environment and the MPLAB C30 language suite. Even if you have never heard of the C language before, you might have heard of the famous "Hello World!" programming example. If not, let me tell you about it.

Since the very first book on the C language, written by Kernighan and Ritchie several decades ago, every decent C-language book has featured an example program containing a single statement to display the words "Hello World" on the computer screen. Hundreds, if not thousands, of books have respected this tradition, and I don't want this book to be the exception. However, it will have to be just a little different. Let's be realistic—we are talking about programming microcontrollers because we want to design embedded-control applications. While the availability of a monitor screen is a perfectly safe assumption for any personal computer or workstation, this is definitely not the case in the embedded-control world. For our first embedded application, we better stick to a more basic type of output—a digital I/O pin. In a later and more advanced chapter, we will be able to interface to an LCD display and/or a terminal connected to a serial port. But by then we will have better things to do than writing "Hello World!"

Preflight checklist

Each flight is preceded by a preflight inspection—simply a walk around the airplane in which we check that, among many other things, gas is in the tank and the wings are still attached to the fuselage. So, let's verify we have all the necessary pieces of equipment ready and installed (from the attached CD-ROM and/or the latest version available for download from Microchip's web site at *http://www.microchip.com/mplab*):

- MPLAB IDE, free Integrated Development Environment
- MPLAB SIM, software simulator
- MPLAB C30, C compiler (free Student Version).

Then, let's follow the "New Project Set-up" checklist to create a new project with the MPLAB IDE:

1. Select "Project→Project Wizard" to activate the new project wizard, which will guide us automatically through the following steps...

2. Select the PIC24FJ128GA010 device, and click Next.

3. Select the MPLAB C30 Compiler Suite and click Next.

4. Create a new folder and name it "Hello"; name the project "Hello Embedded World" and click Next.

5. Simply click Next to the following dialog box—there is no need to copy any source files from any previous projects or directories.

6. Click on Finish to complete the Wizard set-up.

For this first time, let's continue with the following additional steps:

7. Open a new editor window.

8. Type the following three comment lines:

```
//
//    Hello Embedded World!
//
```

9. Select "File→Save As", to save the file as: "Hello.c".

10. Select "Project→Save" to save the project.

The flight

It is time to start writing some code. I can see your trepidation, especially if you have never written any C code for an embedded-control application before. Our first line of code is going to be:

```
#include <p24fj128ga010.h>
```

This is not yet a proper C statement, but more of a pseudo-instruction for the preprocessor telling the compiler to read the content of a device-specific file before proceeding any further. The content of the device-specific ".h" file chosen is nothing more than a long list of the names (and sizes) of all the internal special-function registers (SFRs) of the chosen PIC24 model. If the include file is accurate, those names reflect exactly those being used in the device datasheet. If any doubt, just open the file and take a look—it is a simple text file you can open with the MPLAB editor. Here is a segment of the p24fj128ga010.h file where the program counter and a few other special-function registers (SFRs) are defined:

```
...
extern volatile unsigned int  PCL __attribute__((__sfr__));
extern volatile unsigned char PCH __attribute__((__sfr__));
extern volatile unsigned char TBLPAG __attribute__((__sfr__));
extern volatile unsigned char PSVPAG __attribute__((__sfr__));
extern volatile unsigned int  RCOUNT __attribute__((__sfr__));
extern volatile unsigned int  SR __attribute__((__sfr__));
...
```

Going back to our "Hello.c" source file, let's add a couple more lines that will introduce you to the main() function:

```
main()
{

}
```

What we have now is already a complete, although still empty and pretty useless, C-language program. In between those two curly brackets is where we will soon put the first few instructions of our embedded-control application.

Independently of this function position in the file, whether in the first lines on top or the last few lines in a million-line file, the main() function is the place where the microcontroller (program counter) will go first at power-up or after each subsequent reset.

One caveat—before entering the main() function, the microcontroller will execute a short initialization code segment automatically inserted by the linker. This is known as the c0 code. The c0 code will perform basic housekeeping chores, including the initialization of the microcontroller stack, among other things.

We said our mission was to turn on one or more I/O pins: say PORTA, pins RA0–7. In assembly, we would have used a pair of mov instructions to transfer a literal value to the output port. In C it is much easier—we can write an "assignment statement" as in the following example:

```
#include <p24fj128ga010.h>

main()
{
    PORTA = 0xff;
}
```

First, notice how each individual statement in C is terminated with a semicolon. Then notice how it resembles a mathematical equation...it is not!

An assignment statement has a right side, which is computed first. A resulting value is obtained (in this case it was simply a literal constant) and it is then transferred to the left side, which acts as a receiving container. In this case it was a special-function 16-bit register of the microcontroller (the name of which was predefined in the `.h` file).

Note: In C language, by prefixing the literal value with `0x`, we indicate the use of the hexadecimal radix. Otherwise the compiler assumes the default decimal radix. Alternatively, the `0b` prefix can be used for binary literal values, while for historical reasons a single `0` (zero) prefix is used for the octal notation. (Does anybody use octal anymore?)

Compiling and linking

Now that we have completed the `main()` and only function of our first C program, how do we transform the source into a binary executable?

Using the MPLAB Integrated Development Environment (IDE), it is very easy! It's a matter of a single click of your mouse. This operation is called a Project Build. The sequence of events is fairly long and complex, but it is composed mainly of two steps:

- Compiling: The C compiler is invoked and an object code file (`.o`) is generated. This file is not yet a complete executable. While most of the code generation is complete, all the addresses of functions and variables are still undefined. In fact, this is also called a relocatable code object. If there are multiple source files, this step is repeated for each one of them.

- Linking: The linker is invoked and a proper position in the memory space is found for each function and each variable. Also any number of precompiler object code files and standard library functions may be added at this time as required. Among the several output files produced by the linker is the actual binary executable file (`.hex`).

All this is performed in a very rapid sequence as soon as you select the option "Build All" from the Project menu.

Should you prefer a command-line interface, you will be pleased to learn that there are alternative methods to invoke the compiler and linker and achieve the same results without using the MPAB IDE, although you will have to refer to the MPLAB C compiler User Guide for instructions. In the remainder of this book, we will stick to the MPLAB IDE interface and we will make use of the appropriate checklists to make it even easier.

In order for MPLAB to know which file(s) need to be compiled, we will need to add their names (`Hello.c` in this case) to the project Source Files List.

In order for the linker to assign the correct addresses to each variable and function, we will need to provide MPLAB with the name of a device-specific "linker script" file (`.gld`). Just like the include (`.h`) file tells the compiler about the names (and sizes) of device-specific, special-function registers (SFRs), the linker scripts (`.gld`) file informs the linker about their predefined positions in memory (according to the device datasheet) as well as provides essential memory space information such as: total amount of Flash memory available, total amount of RAM memory available, and their address ranges.

The linker script file is a simple text file and it can be opened and inspected using the MPLAB editor.

Here is a segment of the `p24fj128ga010.gld` file where the addresses of the program counter and a few other special-function registers are defined:

```
   . . .
   PCL        = 0x2E;
   _PCL       = 0x2E;
   PCH        = 0x30;
   _PCH       = 0x30;
   TBLPAG     = 0x32;
   _TBLPAG    = 0x32;
   PSVPAG     = 0x34;
   _PSVPAG    = 0x34;
   RCOUNT     = 0x36;
   _RCOUNT    = 0x36;
   SR         = 0x42;
   _SR        = 0x42;
   . . .
```

Building the first project

Let's review the last few steps required to complete our first demo project:

1. Add the current source file to the "Project Source Files" list.
    There are three possible checklists to choose from, corresponding to three different methods to achieve the same result. This first time we will:
    a) Open the Project window, if not already open, selecting "View→Project".
    b) With the cursor on the editor window, right click to activate the editor pop-up menu.
    c) Select "Add to project".

2. Add the PIC24 "linker script" file to the Project.
    Following the appropriate checklist "Add linker script to project":
    a) Right click on the linker scripts list in the project window.
    b) Select "Add file," browse and select the "`p24fj128ga010.gld`" file found in the `support/gld` subdirectory of MPLAB.

Your Project window should now look similar to Figure 1-1.

3. Select the "Project→Build" function and watch the C30 compiler, followed by the linker, work and generate the executable code as well as a few, hopefully reassuring, messages in the MPLAB IDE Build window.

> Note: The "Project Build" checklist contains several additional steps that will help you in future and more complex examples. (See Figure 1-2.)

4. Select "Debugger→Select Tool→MPLAB SIM" to select and activate the simulator as our main debugging tool for this lesson. Note: the "MPLAB SIM debugger set-up" checklist will help you properly configure the simulator.

If all is well, before trying to run the code, let's also open a Watch window and select and add the PORTA special-function register to it (type or select PORTA in the SFR combo box, and then click on the "Add SFR" button). (See Figure 1-3.)

Figure 1-1. MPLAB IDE Project window set up for the "Hello Embedded World" project.

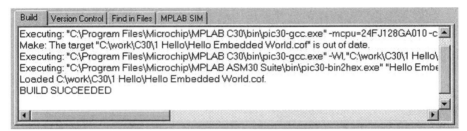

Figure 1-2. MPLAB IDE Output window, Build tab after successfully building a project.

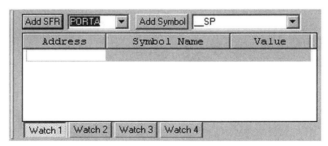

Figure 1-3. MPLAB IDE Watch window.

Figure 1-4. MPLAB IDE Editor context menu (right click).

5. Hit the simulator Reset button 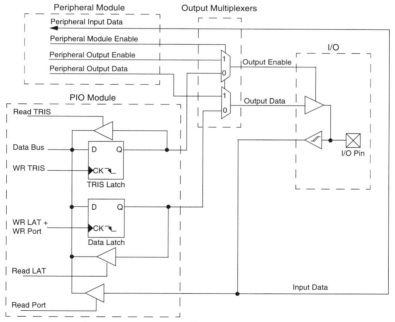 (or select "Debugger→Reset") and observe the contents of PORTA. It should be cleared at reset. Then, place the cursor on the line containing the port assignment, inside the main function, and select the "Run to Cursor" option on the right-click menu.

This will let you skip all the C-compiler initialization code (c0) and get right to the beginning of our code.

6. Now single-step, (use the Step-Over ⟨icon⟩ or Step-In ⟨icon⟩ functions) to execute the one and only statement in our first program and observe how the content of PORTA changes in the Watch window. Or, notice how nothing happens: surprise!

PORT initialization

It is time to hit the books, specifically the PIC24FJ128GA datasheet (Chapter 9, for the I/O ports detail). PORTA is a pretty busy, 16-pin wide, port.

Figure 1-5. Diagram of a typical PIC24 I/O port.

Looking at the pin-out diagrams on the datasheet, we can tell there are many peripheral modules being multiplexed on top of each pin. We can also determine what the default direction is for all I/O pins at reset: they are configured as inputs, which is a standard for all PIC® microcontrollers. The TRISA special-function register controls the direction of each pin on PORTA. Hence, we need to add one more assignment to our program, to change the direction of all the pins of PORTA to output, if we want to see their status change:

```
#include <p24fj128ga010.h>
```

```
main()
{
    TRISA = 0;              // all PORTA pins output
    PORTA = 0xff;
}
```

Retesting PORTA

1. Rebuild the project now.

2. Set the cursor on the TRISA assignment.

3. Execute a "Run to Cursor" command to skip all the compiler initialization just as we did before.

4. Execute a couple of single steps and...we have it!

Address	Symbol Name	Value
02C2	PORTA	0x00FF

Figure 1-6. MPLAB IDE Watch window detail; PORTA content has changed!

If all went well, you should see the content of PORTA change to 0x00FF, highlighted in the Watch window in red. Hello, World!

Our first choice of PORTA was dictated partially by the alphabetical order and partially by the fact that, on the popular Explorer16 demonstration boards, PORTA pins RA0 through RA7 are conveniently connected to 8 LEDs. So if you try to execute this example code on the actual demo board, you will have the satisfaction of seeing all the LEDs turn on, nice and bright.

Testing PORTB

To complete our lesson, we will now explore the use of one more I/O port, PORTB.

It is simple to edit the program and replace the two PORTA control register assignments with TRISB and PORTB. Rebuild the project and follow the same steps we did in the previous exercise and...you'll get a new surprise. The same code that worked for PORTA does not work for PORTB!

Don't panic! I did it on purpose. I wanted you to experience a little PIC24 migration pain. It will help you learn and grow stronger.

It is time to go back to the datasheet, and study in more detail the PIC24 pin-out diagrams. There are two fundamental differences between the 8-bit PIC microcontroller architecture and the new PIC24 architecture:

- Most of PORTB pins are multiplexed with the analog inputs of the analog-to-digital converter (ADC) peripheral. The 8-bit architecture reserved PORTA pins primarily for this purpose—the roles of the two ports have been swapped!

- With the PIC24, if a peripheral module input/output signal is multiplexed on an I/O pin, as soon as the module is enabled, it takes complete control of the I/O pin—independently of the direction (TRISx) control register content. In the 8-bit architectures, it was up to the user to assign the correct direction to each pin, even when a peripheral module required its use.

By default, pins multiplexed with "analog" inputs are disconnected from their "digital"input ports. This is exactly what was happening in the last example. All PORTB pins in the PIC24FJ128GA010 are, by default at power-up, assigned an analog input function; therefore, reading PORTB returns all 0s. Notice, though, that the output latch of PORTB has been correctly set although we cannot see it through the PORTB register. To verify it, check the contents of the LATB register instead.

To reconnect PORTB inputs to the digital inputs, we have to act on the analog-to-digital conversion (ADC) module inputs. From the datasheet, we learn that the special-function register AD1PCFG controls the analog/digital assignment of each pin.

Upper Byte:

R/W-0	R/W-0	R/W-0	R/W-0	R/W-0	R/W-0	R/W-0	R/W-0
PCFG15	PCFG14	PCFG13	PCFG12	PCFG11	PCFG10	PCFG9	PCFG8
bit 15							bit 8

Lower Byte:

R/W-0	R/W-0	R/W-0	R/W-0	R/W-0	R/W-0	R/W-0	R/W-0
PCFG7	PCFG6	PCFG5	PCFG4	PCFG3	PCFG2	PCFG1	PCFG0
bit 7							bit 0

bit 15-0 **PCFG15:PCFG0:** Analog Input Pin Configuration Control bits
    1 = Pin for corresponding analog channel is configured in Digital mode; I/O port read enabled
    0 = Pin configured in Analog mode; I/O port read disabled, A/D samples pin voltage

Figure 1-7. AD1PCFG: *ADC port configuration register.*

Assigning a 1 to each bit in the AD1PCGF special-function register will accomplish the task. Our new and complete program example is now:

```
#include <p24fj128ga010.h>

main()
{
    TRISB =        0;           // all PORTB pins output
    AD1PCFG =      0xffff;      // all PORTB pins digital
    PORTB =        0xff;
}
```

This time, compiling and single-stepping through it will give us the desired results.

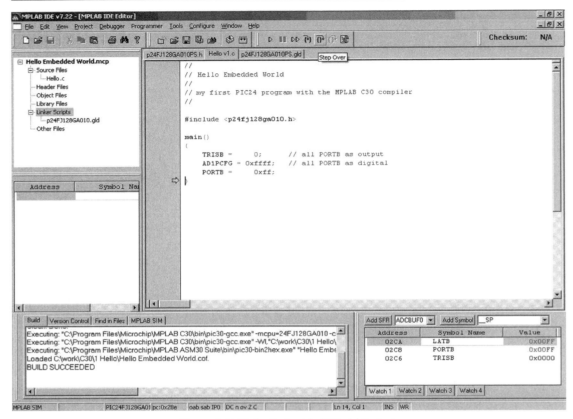

Figure 1-8. Hello Embedded World Project.

Post-flight briefing

After each flight, there should be a brief review. Sitting on a comfortable chair in front of a cool glass of water, it's time to reflect with the instructor on what we have learned from this first experience.

Writing a C program for a PIC24 microcontroller can be very simple, or at least no more complicated than the assembly-language equivalent. Two or three instructions, depending on which port we plan to use, can give us direct control over the most basic tool available to the microcontroller for communication with the rest of the world: the I/O pins.

Also, there is nothing the C30 compiler can do to read our mind. Just like in assembly, we are responsible for setting the correct direction of the I/O pins. And we are still required to study the datasheet and learn about the small differences between the 8-bit PIC microcontrollers we might be familiar with and the new 16-bit breed.

As high-level as the C programming language is touted to be, writing code for an embedded-control device still requires us to be intimately familiar with the finest details of the hardware we use.

Notes for assembly experts

If you have difficulties blindly accepting the validity of the code generated by the MPLAB C30 compiler, you might find comfort in knowing that, at any given point in time, you can decide to switch to the "Disassembly Listing" view. You can quickly inspect the code generated by the compiler, as each C source line is shown in a comment that precedes the segment of code it generated.

```
■ Disassembly Listing                                               _ □ ×
  6:
  7:                     #include <p24fj128ga010.h>
  8:
  9:                     main()
 10:                     {
  00280  FA0000    lnk #0x0
 11:                       TRISA =        0;      // all PORTA as output
  00282  EB0000    clr.w 0x0000
  00284  881600    mov.w 0x0000,0x02c0
 12: ·                     AD1PCFG = 0xffff;    // all PORTA as digital
  00286  EB8000    setm.w 0x0000
  00288  881960    mov.w 0x0000,0x032c
 13:                       PORTA =       0xff;
  0028A  200FF0    mov.w #0xff,0x0000
  0028C  881610    mov.w 0x0000,0x02c2
 14:                     }
  0028E  FA8000    ulnk
  00290  060000    return
```

Figure 1-9. Disassembly Listing Window.

You can even single-step through the code and do all the debugging from this view, although I strongly encourage you not to do so (or limit the exercise to a few exploratory sessions as we progress through the first chapters of this book). Satisfy your curiosity, but gradually learn to trust the compiler. Eventually, use of the C language will give a boost to your productivity and increase the readability and maintainability of your code.

As a final exercise, I encourage you to open the Memory Usage Gauge window—select "View→ Memory Usage Gauge".

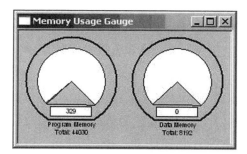

Figure 1-10. MPLAB IDE Memory Usage Gauge window.

Don't be alarmed! Even though we wrote only three lines of code in our first example and the amount of program memory used appears to already be up to 300+ bytes, this is not an indication of any inherent inefficiency of the C language. There is a minimum block of code that is always generated (for our convenience) by the C30 compiler. This is the initialization code (c0) that we mentioned briefly before. We will get to it, in more detail, in the following chapters as we discuss variables initialization, memory allocation and interrupts.

Notes for PIC MCU experts

Those of you who are familiar with the PIC16 and PIC18 architecture will find it interesting that most PIC24 control registers, including the I/O ports, are now 16 bits wide. Looking at the PIC24 datasheet, note also how most peripherals have names that look very similar, if not identical, to the 8-bit peripherals you are already familiar with. You will feel at home in no time!

Notes for C experts

Certainly we could have used the `printf` function from the standard C libraries. In fact the libraries are readily available with the MPLAB C30 compiler. But we are targeting embedded-control applications and we are not writing code for multigigabyte workstations. Get used to manipulating low-level hardware peripherals inside the PIC24 microcontrollers. A single call to a library function, like `printf`, could have added several kilobytes of code to your executable. Don't assume a serial port and a terminal or a text display will always be available to you. Instead, develop a sensibility for the "weight" of each function and library you use in light of the limited resources available in the embedded-design world.

Tips and tricks

The PIC24FJ family of microcontrollers is based on a 3V CMOS process with a 2.0V to 3.6V operating range. As a consequence, a 3V power supply (Vdd) must be used and this limits the output voltage of each I/O pin when producing a logic high output. However, interfacing to 5V legacy devices and applications is really simple:

*   To drive a 5V output, use the ODCx control registers (ODCA for PORTA, ODCB for PORTB and so on...) to set individual output pins in open-drain mode and connect external pull-up resistors to a 5V power supply.

*   Digital input pins instead are already capable of tolerating up to 5V. They can be connected directly to 5V input signals.

Be careful with I/O pins that are multiplexed with analog inputs though—they cannot tolerate voltages above Vdd.

Exercises

If you have the Explorer16 board:

1. Use the ICD2 Debugging Checklist to help you prepare the project for debugging.

2. To test the PORTA example, connect the Explorer16 board and check the visual output on LED0–7.

3. To test the PORTB example, connect a voltmeter (or DMM) to pin RB0 and watch the needle move as you single-step through the code.

Books

*   Kernighan, B. and Ritchie, D., "**The C Programming Language**," Prentice-Hall, Englewood Cliffs, NJ.

    When you read or hear a programmer talk about the "K&R," they mean this book! Also known as "the white book," the C language has evolved since the first edition of this book was published in 1978! The second edition (1988) includes the more recent ANSI C standard definitions of the language, which is closer to the standard the MPLAB C30 compiler adheres to (ANSI90).

*   "**Private Pilot Manual**," Jeppesen Sanderson, Inc., Englewood, CO.

    This is "the" reference book for every student pilot. Highly recommended, even if you are just curious about aviation.

Links

*   *http://en.wikibooks.org/wiki/C_Programming*

    This is a Wiki-book on C programming. It's convenient if you don't mind doing all your reading online. Hint: look for the chapter called "A taste of C" to find the omnipresent "Hello World!" exercise.

A loop in the pattern

In This Chapter

▶ *while loops*
▶ *An animated simulation*
▶ *Using the Logic Analyzer*

The "pattern" is a standardized rectangular circuit, where each pilot flies in a loop. Every airport has a pattern of given (published) altitude and position for each runway. Its purpose is to organize traffic around the airport and its working is not too dissimilar to how a roundabout works. All airplanes are supposed to circle in a given direction consistent with the prevailing wind at the moment. They all fly at the same altitude, so it is easier to visually keep track of each other's position. They all talk on the radio on the same frequencies, communicating with a tower if there is one, or among one another with the smaller airports. As a student pilot, you will spend quite some time, especially in the first few lessons, flying in the pattern with your instructor to practice continuous sequences of landings immediately followed by take-offs (touch-and-goes), refining your newly acquired skills. As a student of embedded programming, you will have a loop of your own to learn—the main loop.

Flight plan

Embedded-control programs need a framework, similar to the pilots' pattern, so that the flow of code can be managed. In this lesson, we will review the basics of the loops syntax in C and we'll also take the opportunity to introduce a new peripheral module: the 16-bit Timer1. Two new MPLAB® SIM features will be used for the first time: the "Animate" mode and the "Logic Analyzer."

Preflight checklist

For this second lesson, we will need the same basic software components installed (from the attached CD-ROM and/or the latest versions, available for download from Microchip's website) and used before, including:

* MPLAB IDE, Integrated Development Environment

* MPLAB SIM, software simulator

* MPLAB C30 compiler (Student Version)

We will also reuse the "New Project Set-up" checklist to create a new project with the MPLAB IDE:

1. Select "Project→Project Wizard", to start creating a new project.

2. Select the PIC24FJ128GA010 device, and click Next.

3. Select the MPLAB C30 compiler suite and click Next.

4. Create a new folder and name it "Loop." name the project "A Loop in the Pattern," and click Next.

5. There is no need to copy any source files from the previous lessons; click Next once more.

6. Click Finish to complete the Wizard set-up.

This will be followed by the "Adding Linker Script file" checklist, to add the linker script "p24fj128ga010.gld" to the project. It can typically be found in the MPLAB IDE installation directory "C:/Program Files/Microchip/", within the subdirectory "MPLAB C30/support/gld/".

After completing the "Create New File and Add to Project" checklist:

7. Open a new editor window.

8. Type the main program header:

```
//
//    A loop in the pattern
//
```

9. Select "Project→AddNewFiletoProject", to save the file as: "loop.c" and have it automatically added to the project source files list.

10. Save the project.

The flight

One of the key questions that might have come to mind after working through the previous lesson is "What happens when all the code in the main() function has been executed?" Well, nothing really happens, or at least nothing that you would not expect. The device will reset, and the entire program will execute again…and again.

In fact, the compiler puts a special software reset instruction right after the end of the main() function code, just to make sure. In embedded control we want the application to run continuously, from the moment the power switch has been flipped on until the moment it is turned off. So, letting the program run through entirely, reset and execute again might seem like a convenient way to arrange the application so that it keeps repeating as long as there is "juice." This option might work in a few limited cases, but what you will soon discover is that, running in this "loop." you develop a "limp." Reaching the end of the program and executing a reset takes the microcontroller back to the very beginning to execute all the initialization code, including the c0 code segment briefly mentioned in the previous lesson. So, as short as the initialization part might be, it will make the loop very unbalanced. Going through all the special-function register and global-variables initializations each time is probably not necessary and it will certainly slow down the application. A better option is to design an application "main loop." Let's review the most basic loop-coding options in C first.

While **loops**

In C there are at least three ways to code a loop; here is the first—the `while` loop:

```
while ( x)
{
    // your code here...
}
```

Anything you put between those two curly brackets ({ }) will be repeated for as long as the logic expression in parenthesis (x) returns a true value. But what is a logic expression in C?

First of all, in C there is no distinction between logic expressions and arithmetic expressions. In C, the Boolean logic TRUE and FALSE values are represented just as integer numbers with a simple rule:

*   FALSE is represented by the integer 0.

*   TRUE is represented by *any* integer except 0.

So 1 is true, but so are 13 and -278. In order to evaluate logic expressions, a number of logic operators are defined, such as:

||	the logic OR operator,
&&	the logic AND operator,
!	the logic NOT operator.

These operators consider their operands as logical (Boolean) values using the rule mentioned above, and they return a logical value. Here are some trivial examples:

(when a = 17 and b = 1, or in other words they are both true)

(a || b)	is true,
(a && b)	is true
(!a)	is false

There are, then, a number of operators that compare numbers (integers of any kind and floating-point values, too) and return logic values. They are:

==   the "equal-to" operator; notice it is composed of two equal signs to distinguish it from the "assignment" operator we used in the previous lesson,

!=   the "NOT-equal to" operator,

>   the "greater-than" operator,

>=   the "greater-or-equal to" operator,

<   the "less-than" operator,

<=   the "less-than-or-equal to" operator.

Here are some examples:

assuming a = 10

(a > 1)	is true
(-a >= 0)	is false

```
( a == 17) is false
( a != 3)   is true
```

Back to the `while` loop, we said that as long as the expression in parentheses produces a true logic value (that is any integer value but 0), the program execution will continue around the loop. When the expression produces a false logic value, the loop will terminate and the execution will continue from the first instruction after the closing curly bracket.

Note that the expression is evaluated first, before the curly bracket content is executed (if ever), and is then reevaluated each time.

Here are a few curious loop examples to consider:

```
While ( 0)
{
    // your code here...
}
```

A constant false condition means that the loop will never be executed. This is not very useful. In fact I believe we have a good candidate for the "world's most useless code" contest!

Here is another example:

```
while ( 1)
{
    // your code here...
}
```

A constant true condition means that the loop will execute forever. This is useful, and is in fact what we will use for our main program loops from now on. For the sake of readability, a few purists among you will consider using a more elegant approach, defining a couple of constants:

```
#define TRUE      1
#define FALSE     0
```

and using them consistently in their code, as in:

```
While ( TRUE)
{
    // your code here…
}
```

It is time to add a few new lines of code to the "`loop.c`" source file now, and put the `while` loop to good use.

```
#include <p24fj128ga010.h>
```

```
main()
{
    // init the control registers
    TRISA =    0xff00;// PORTA pin 0..7 as output

    // application main loop
    while( 1)
    {
        PORTA = 0xff; // turn pin 0-7 on
        PORTA = 0;              // turn all pin off
    }
}
```

The structure of this example program is essentially the structure of every embedded-control program written in C. There will always be two main parts:

- The initialization, which includes both the device peripherals initialization and variables initialization, executed only once at the beginning.

- The main loop, which contains all the control functions that define the application behavior, and is executed continuously.

An animated simulation

Use the Project Build checklist to compile and link the "loop.c" program. Also use the "MPLAB SIM simulator set-up" checklist to prepare the software simulator.

To test the code in this example with the simulator, I recommend you use the "Animate" mode (Debugger→Animate). In this mode, the simulator executes one C program line at a time, pausing for ½ second after each one to give us the time to observe the immediate results. If you add the PORTA special-function register to the Watch window, you should be able to see its value alternating rhythmically between 0xff and 0x00.

The speed of execution in Animate mode can be controlled with the "Debug→Settings" dialog box, selecting the "Animation/Real Time Updates" tab, and modifying the "Animation Step Time" parameter, which by default is set to 500 ms. As you can imagine, the "Animate" mode can be a valuable and entertaining debugging tool, but it gives you quite a distorted idea of what the actual program execution timing will be. In practice, if our example code was to be executed on a real hardware target, say an Explorer16 demonstration board (where the PIC24 is running at 32 MHz), the LEDs connected to the PORTA output pins would blink too fast for our eyes to notice. In fact, each LED would be turned on and off several million times each second.

To slow things down to a point where the LEDs would blink nicely just a couple of times per second, I propose we use a timer, so that in the process we learn to use one of the key peripherals integrated in all PIC24 microcontrollers. For this example, we will choose the first timer, Timer1, of the five timers available inside the PIC24FJ128GA010. This is one of the most flexible and simple peripheral modules. All we need to do is take a quick look at the PIC24 datasheet, check the block diagram and the details of the Timer1 control registers, and find the ideal initialization values.

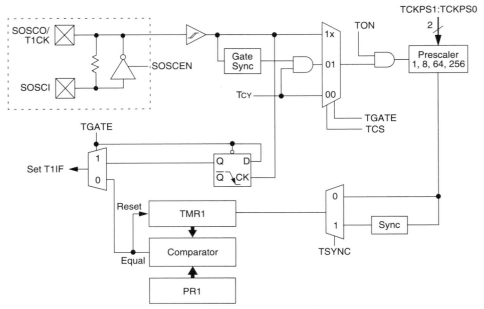

Figure 2-1. 16-bit Timer1 Module block diagram.

We quickly learn that there are three special-function registers that control most of the Timer1 functions. They are:

- `TMR1`, which contains the 16-bit counter value.

- `T1CON`, which controls activation and the operating mode of the timer.

- `PR1`, which can be used to produce a periodic reset of the timer
  (not required here).

We can clear the `TMR1` register to start counting from zero.

```
TMR1 = 0;
```

Then we can initialize `T1CON` so that the timer will operate in a simple configuration where:

- Timer1 is activated: `TON = 1`

- The main MCU clock serves as the source (Fosc/2): `TCS = 0`

- The prescaler is set to the maximum value (1:256): `TCKPS = 11`

- The input gating and synchronization functions are not required, since we use the MCU internal clock directly as the timer clock: `TGATE = 0, TSYNC = 0`

- We do not worry about the behavior in IDLE mode: `TSIDL = 0 (default)`

Upper Byte:							
R/W-0	U-0	R/W-0	U-0	U-0	U-0	U-0	U-0
TON	—	TSIDL	—	—	—	—	—
bit 15							bit 8

Lower Byte:							
U-0	R/W-0	R/W-0	R/W-0	U-0	R/W-0	R/W-0	U-0
—	TGATE	TCKPS1	TCKPS0	—	TSYNC	TCS	—
bit 7							bit 0

Figure 2-2. T1CON: Timer1 control register.

Once we assemble all the bits in a single 16-bit value to assign to T1CON, we get:

```
T1CON = 0b1000000000110000;
```

or, in a more compact hexadecimal notation:

```
T1CON = 0x8030;
```

Once we are done initializing the timer, we enter a loop where we wait for TMR1 to reach the desired value set by the constant DELAY.

```
while( TMR1 < DELAY)
{
    // wait
}
```

Assuming a 32-MHz clock will be used, we need to assign quite a large value to DELAY so as to obtain a delay of about a quarter of a second. In fact the following formula dictates the total delay time produced by the TMR1 loop:

$$Tdelay = (2/Fosc) * 256 * DELAY$$

With Tdelay = 256 ms and resolving for DELAY, we obtain the value 16,000:

```
#define DELAY 16000
```

By putting two such delay loops in front of each PORTA assignment inside the main loop, we get our latest and best code example:

```
#include <p24fj128ga010.h>

#define DELAY      16000

main()
{
    // init the control registers
    TRISA = 0xff00;            // PORTA pin 0..7 as output
    T1CON = 0x8030;        // TMR1 on, prescaler 1:256 Tclk/2

    // main application loop
    while( 1)
    {
```

```
                    // 1. turn pin 0-7 on and wait for ¼ second
                    PORTA = 0xff;
                    TMR1 = 0;        // restart the count
                    while ( TMR1 < DELAY)
                    {
                            // just wait
                    }

                    // 2. turn all pin off and wait for ¼ second
                    PORTA = 0x00;
                    TMR1 = 0;        // restart the count
                    while ( TMR1 < DELAY)
                    {
                            // just wait
                    }

            } // main loop
    } // main
```

Note: When programming in C, the number of opening and closing curly brackets tends to increase rapidly as your code grows. After a very short while, even if you stick religiously to the best indentation rules, it can become difficult to remember which closing curly brackets belong to which opening curly brackets. By putting little reminders (comments) on the closing brackets, I try to make it easier and more readable.

It is time now to build the complete project and verify that it is working. If you have an Explorer16 demonstration board available, you may try to run the code right away. The LEDs should flash at a comfortably slow pace, with a frequency of about two flashes per second.

If you try to run the same code with the MPLAB SIM simulator, though, you will discover that things are now way too slow. I don't know how fast your PC is, but on mine MPLAB-SIM cannot get anywhere close to the execution speed of a true 32-MHz PIC24 microcontroller.

If you use the Animate mode, things get even worse. As we saw before, the animation adds a further delay of about half a second between the execution of each individual line of code. So, for pure debugging purposes, on the simulator feel free to change the DELAY constant to a much smaller value (16, for example).

Using the Logic Analyzer

To complete this lesson and make things more entertaining, after building the project, I suggest we play with a new simulation tool: the MPLAB logic analyzer.

The logic analyzer gives you a graphical and extremely effective view of the recorded values for any number of the device output pins, but it requires a little care in the initial set-up.

Before anything else, you should make sure that the Tracing function of the simulator is turned on.

1. Select the "Debug→Settings" dialog box and then choose the Osc/Trace tab.

2. In the Tracing options section, check the Trace All box.

3. Now you can open the Analyzer window, from the "View→Simulator" Logic Analyzer menu.

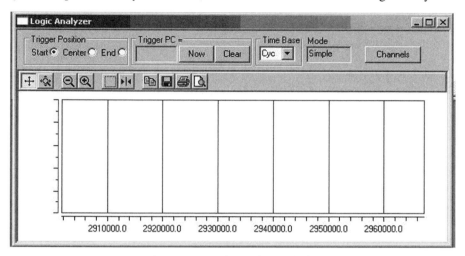

Figure 2-3. Logic Analyzer window.

4. Then click on the channel button, to bring up the channel-selection dialog box.

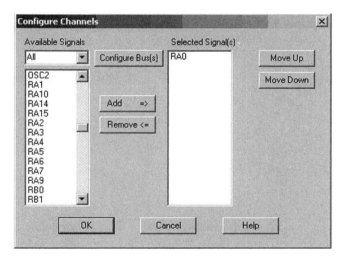

Figure 2-4. Channel Selection Dialog Box.

5. From here, you can select the device output pins you would like to visualize. In our case, select RA0 and click "Add =>".

6. Click on OK to close the channel-selection dialog box.

Note: For future reference, all the steps above are listed in the "Logic Analyzer Set-up" checklist.

Run the code ▷ for a short while and then hit the Halt button ⎕⎕. The Logic Analyzer window should display a neat square-wave plot, as in Figure 2-5.

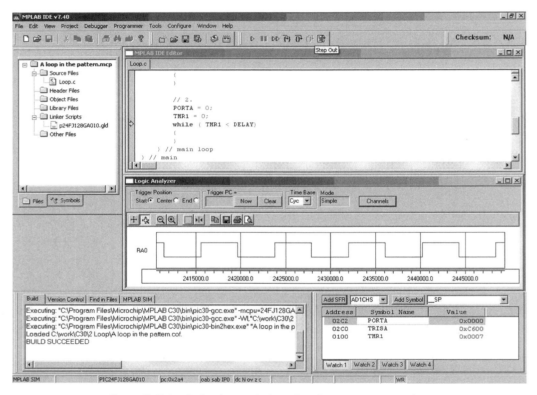

Figure 2-5. Logic Analyzer window showing square-wave plot.

Post-flight briefing

In this brief lesson, we learned about the way the MPLAB C30 compiler deals with program termination. For the first time, we gave our little project a bit of structure—separating the main() function in an initialization section and an infinite loop. To do so, we learned about the while loop statements and we took the opportunity to touch briefly on the subject of logical expressions evaluation. We closed the lesson with a final example, where we used a timer module for the first time and we played with the Logic Analyzer window to plot the RA0 pin output.

We will return to all these elements, so don't worry if you have more doubts now than when we started—this is all part of the learning experience.

Notes for assembly experts

Logic expressions in C can be tricky for the assembly programmer who is used to dealing with *binary* operators of identical names (AND, OR, NOT…). C has a set of binary operators too, but I purposely avoided showing them in this lesson to avoid mixing things up. Binary logic operators take pairs of bits from each operand and compute the result according to the defined truth table. Logic operators, on the other hand, look at each operand (independently of the number of bits used) as a single Boolean value.

See the following examples of byte-sized operands:

```
            11110101                    11110101  (TRUE)
binary OR   00001000      logical OR    00001000  (TRUE)
            --------                    --------
gives       11111101      gives         00000001  (TRUE)
```

Notes for PIC microcontroller experts

I am sure you noticed: Timer0 has disappeared! The good news is: you are not going to miss it! In fact, the remaining five timers of a PIC24 are so loaded with features that there is no functionality of Timer0 that you are going to feel nostalgic about. All of the special-function registers that control the timers have names similar to the ones used on PIC16 and PIC18 microcontrollers, and are pretty much identical in structure. Still, keep an eye on the datasheet; the designers managed to cram in several new features, including:

- All timers are now 16 bits wide.
- Each timer has a 16-bit period register.
- A new 32-bit mode timer-pairing mechanism is available for Timer2/3 and Timer4/5.
- A new external clock gating feature has been added on Timer1.

Notes for C experts

If you are used to programming in C on a personal computer or workstation, you expect that, upon termination of the `main()` function, control would be returned to the operating system. While several real-time operating systems (RTOSs) are available for the PIC24, a large number of applications don't need and won't use one. This is certainly true for all the simple examples in this book. By default, the C30 compiler assumes there is no operating system to return control to, and does the safest possible thing—it resets.

Tips and tricks

Some embedded applications are designed to run their main loop for months or years in a row without ever being turned off or receiving a reset command. But the control registers of a microcontroller are simple RAM memory cells. The probability that a power-supply fluctuation (undetected by the brown-out reset circuit), an electromagnetic pulse emitted by some noisy equipment in the proximity, or even a cosmic ray could alter their contents is a small but finite number. Given enough time, depending on the application, you will see it happen. When you design applications that have to operate reliably on such huge time scales, you should start seriously considering the need to provide a periodic "refresh" of the most important control registers of the essential peripherals used by the application.

Group the sequence of initialization instructions in one or more functions. Call the functions once at power-up, before entering the main loop, but also make sure that inside the main loop the initialization functions are called when no other critical task is pending and every control register is reinitialized periodically.

Exercises

1. Output a counter on the PORTA pins instead of the alternating on and off patterns.

2. Use a rotating pattern instead of alternating on and off.

Books

• Ullman, L. and Liyanage, M. (2005)

   C Programming

   Peachpit Press, Berkeley, CA.

   This is a fast-reading and modern book, with a simple step-by-step introduction to the C programming language.

• Adams, N. (2003)

   The Flyers, in Search of Wilbur and Orville Wright

   Three Rivers Press, New York, NY

   A trip back in time to the first powered flight in history, just 120 feet on the sands of Kitty Hawk.

Links

• *http://en.wikipedia.org/wiki/Control_flow#Loops*

   A wide perspective on programming languages and the problems related to coding and taming loops.

• *http://en.wikipedia.org/wiki/Spaghetti_code*

   Your code gets out of control when you cannot fly the pattern.

More pattern work, more loops

In aviation, a proper "loop" is an "aerobatic" maneuver performed only by pilots that have received advanced training, using airplanes that are specially equipped for the task. You could take this as either a disappointment or a reassurance, but you can be certain that, when preparing for the private-pilot license, you will *not* be asked to perform any such trick. There will be plenty of other challenges, though, as you will be asked to perform and repeat to perfection a variety of "turns" including: turns around a point, S turns, steep turns and standard rate turns. In all these exercises, you will discover how difficult it can be—while navigating in a three-dimensional environment—to control only one of the dimensions at a time. When circling around a reference point on the ground, you will initially struggle to maintain a constant altitude and a constant speed. A little bit of wind will add to the challenge of maintaining a constant distance from the reference point, and performing a nice and smooth circle. Practice will make you perfect.

In C-language programming, there are several types of loops, too. Learning which one to choose, and when and why, will take a little practice, but will make you a better embedded-control programmer.

Flight plan

In the previous lesson, we learned there is a loop at the core of every embedded-control application. In this lesson, we will continue exploring a variety of other techniques available to the C programmer to perform loops. Along the way, we will take the opportunity to briefly review integer variables declarations, and increment and decrement operators, quickly touching the arrays declaration and use subject. As in any good flight lesson, the theory is immediately followed by the practice, and we will conclude the lesson with a, hopefully entertaining, exercise that will make use of all the concepts and tools acquired during the lesson.

Preflight checklist

In this lesson we will continue using the MPLAB® SIM software simulator, but once more an Explorer16 demonstration board could add to the entertainment. In preparation for the new demonstration project, you can use the "New Project Set-up" checklist to create a new project called "More Loops" and create a new source file to be called "More.c".

The flight

In a `while` loop, a block of code enclosed by two curly brackets is executed if, and for as long as, a logic expression returns a Boolean true value (not zero). The logic expression is evaluated before the loop, which means that if the expression returns false right from the beginning, the code inside the loop might never be executed.

Do Loops

If you need a type of loop that gets executed at least once, but only subsequent repetitions are dependent on a logic expression, then you have to look at a different type of loop.

Let me introduce you to the `do` loop syntax:

```
do {
    // your code here…


} while ( x );
```

Don't be confused by the fact that the `do` loop syntax is using the `while` keyword again to close the loop—the behavior of the two loop types is very different.

In a `do` loop, the code (if any) found between the curly brackets is always executed first, and only then is the logic expression evaluated. Of course, if all we want is an infinite loop for our `main()` function, then it makes no difference if we choose the `do` or the `while`...

```
main()
{
    // initialization code
    …

    // main application loop
    do {
            …
    } while ( 1 )
} // main
```

Looking for curious cases, we might analyze the behavior of the following loop:

```
do{
    // your code segment here…
} while ( 0 );
```

You will realize that the code segment inside the loop is going to be executed once and, no matter what, only once. In other words, the loop syntax around the code is, in this case, a total waste of your typing efforts and another good candidate for the "most useless piece of code in the world" contest.

Let's now look at a more useful example, where we use a `while` loop to repeatedly execute a piece of code for a predefined and exact number of times. First of all, we need a variable to perform the count. In other words, we need to allocate one or more RAM memory locations to store a counter value.

> Note: In the previous two lessons we have been able to skip almost entirely the subject of variable declarations, as we relied exclusively on the use of what are in fact predefined variables: the special-function registers of the PIC24.

Variable declarations

We can declare an integer variable with the following syntax:

```
int c;
```

Since we used the keyword `int` to declare c as a 16-bit (signed) integer, the MPLAB C30 compiler will make arrangements for two bytes of memory to be used. Later, the linker will determine where those two bytes will be allocated in the physical RAM memory of the selected PIC24 model. As defined, the variable c will allow us to count from a negative minimum value –32,768 to a maximum positive value of +32,767. If we need a larger integer numerical range, we can use the `long` (signed) integer type as in:

```
long c;
```

The MPLAB C30 compiler will use 32 bits (four bytes) for the variable.

If we are looking for a smaller counter, and we can accept a range of values from –128 to +127, we can use the `char` integer type instead:

```
char c;
```

In this case the MPLAB C30 compiler will use 8 bits (a single byte).

All three types can be further modified by the `unsigned` attribute as in:

```
unsigned char c;      // ranges from 0..255
unsigned int  i;      // ranges from 0..65,535
unsigned long l;      // ranges from 0..4,294,967,295
```

There are then variable types defined for use in floating-point arithmetic:

```
float       f;      // defines a 32 bit precision floating point
long double d;      // defines a 64 bit precision floating point variable
```

`for` loops

We can now return to our counter example. All we need is a simple integer variable to be used as index/ counter, capable of covering the range from 0 to 5; therefore a `char` integer type will do:

```
char i;                    // declare i as an 8-bit integer with sign

i = 0;                     // init the index/counter

while ( i<5)
{
```

31

```
        // insert your code here...
        // it will be executed for i= 0, 1, 2, 3, 4

        i = i+1;                 // increment
    }
```

Whether counting up or down, this is something you are going to do a lot in your everyday programming life.

In C language, there is a third type of loop that has been designed specifically to make coding this common case easy. It is called the `for` loop, and this is how you would have used it in the previous example:

```
    for ( i=0; i<5; i=i+1)
    {
        // insert your code here...
        // it will be executed for i=0, 1, 2, 3, 4
    }
```

You will agree that the `for` loop syntax is compact, and it is certainly easier to write. It is also easier to read and debug later. The three expressions separated by semicolons and enclosed in the brackets following the `for` keyword are exactly the same three expressions we used in the prior example:

- initialize the index.

- check for termination, using a logic expression.

- advance the index/counter…in this case incrementing it.

You can think of the `for` loop as an abbreviated syntax of the `while` loop. In fact, the logic expression is evaluated first and, if false from the beginning, the code inside the loop's curly brackets may never be executed.

Perhaps this is also a good time to review another convenient shortcut available in C. There is a special notation reserved for the increment and decrement operations that uses the operators:

| `++` | to increment, as in | `i++;` | is equivalent to | `i = i+1;` |
| `--` | to decrement, as in | `i--;` | is equivalent to | `i = i-1;` |

There will be much more to say on the subject in later chapters, but this will suffice for now.

More loop examples

Let's see some more examples of the use of the `for` loop and the increment/decrement operators.

First, a count from 0 to 4:

```
    for ( i=0; i<5; i++)
    {
        // insert your code here...
        // it will be executed for i= 0, 1, 2, 3, 4
    }
```

Then, a countdown from 4 to 0:

```
for ( i=4; i>=0; i--)
{
   // insert your code here...
   // it will be executed for i= 4, 3, 2, 1, 0
}
```

Can we use the `for` loop to code an (infinite) main program loop?

Sure we can—here is an example:

```
main()
{
   // 0. initialization code
   ...

   // 1. the main application loop
   for ( ; 1; )
   {
         ...
   }
} // main
```

If you like it, feel free to use this form. As for me, from now on I will stick to the `while` syntax (it is just an old habit).

Arrays

Before starting to code our next project, we need to review one last C-language feature: array variable types. An array is just a contiguous block of memory containing a given number of identical elements of the same type. Once the array is defined, each element can be accessed via the array name and an index. Declaring an array is as simple as declaring a single variable—just add the desired number of elements in square brackets after the variable name:

```
char c[10];     // declares c as an array of 10 x 8-bit integers
int i[10];      // declares i as an array of 10 x 16-bit integers
long l[10];     // declares l as an array of 10 x 32-bit integers
```

The same squared-brackets notation is used to refer to the content or assign a value to each element of an array as in:

```
a = c[0];          // copy the value of the 1st element of c into a
c[1] = 123;        // assign the value 123 to the second element of c
i[2] = 12345;      // assign the value 12,345 to the third element of i
l[3] = 123* i[4];  // compute 123 x the value of the fifth element of i
```

Note: In C language, the elements of an array of size N have indexes 0, 1, 2...(N–1). It is when manipulating arrays that the `for` type of loop shows all its merits.

33

Let's see an example where we declare an array of 10 integers and we initialize each element of the array to a constant value of 1:

```
int a[10]; // declare array of 10 integers: a[0], a[1], a[2]…
a[9]
int i;            // the loop index
for ( i=0; i<10; i++)
{
    a[ i] = 1;
}
```

A new demo

The best way to conclude this lesson would be to take all the elements of the C language we have reviewed so far and put them to use in our next project. This project will consist of making a row of LEDs, connected to PORTA (as they happen to be connected on the Explorer16 demo board), flash in a rapid sequence so that when moving the board left and right rhythmically they will display a short text message.

How about "Hello World!" or, perhaps more modestly, "HELLO"?

Here is the code:

```
#include <p24fj128ga010.h>

// 1. define timing constant
#define SHORT_DELAY 100
#define LONG_DELAY 800

// 2. declare and initialize an array with the message bitmap
char bitmap[30] = {
    0b11111111,    // H
    0b00001000,
    0b00001000,
    0b11111111,
    0b00000000,
    0b00000000,
    0b11111111,    // E
    0b10001001,
    0b10001001,
    0b10000001,
    0b00000000,
    0b00000000,
    0b11111111,    // L
    0b10000000,
    0b10000000,
    0b10000000,
    0b00000000,
    0b00000000,
```

```
       0b11111111,   // L
       0b10000000,
       0b10000000,
       0b10000000,
       0b00000000,
       0b00000000,
       0b01111110,   // O
       0b10000001,
       0b10000001,
       0b01111110,
       0b00000000,
       0b00000000
       };

// 3. the main program
main()
{
    // 3.1 variable declarations
    int i;              // i will serve as the index

    // 3.2 initialization
    TRISA =   0xff00;   // PORTA pins connected to LEDs are outputs
    T1CON =   0x8030;   // TMR1 on, prescale 1:256 Tclk/2

    // 3.3 the main loop
    while( 1)
    {
          // 3.3.1 display loop, hand moving to the right
          for( i=0; i<30; i++)
          {       // update the LEDs
                  PORTA = bitmap[i];
                  // short pause
                  TMR1 = 0;
                  while ( TMR1 < SHORT_DELAY)
                  {
                  }
          } // for i

          // 3.3.2 long pause, hand moving back to the left
          PORTA = 0;            // turn LEDs off
          // long pause
          TMR1 =   0;
          while ( TMR1 < LONG_DELAY)
          {
          }
    } // main loop
} // main
```

In section 1, we define a couple of timing constants, so that we can control the flashing sequence speed for execution and debugging.

In section 2, we declare and initialize an 8-bit integer array of 30 elements, each containing an LED configuration in the sequence.

Hint: using a highlighter you can mark the "1s" on the page to see the message emerge.

Section 3 contains the main program, with the variable declarations (3.1) at the top, followed by the microcontroller initialization (3.2) and eventually the main loop (3.3).

The main (`while`) loop, in turn, is divided in two sections:

1.1.1   Contains the actual LED flash sequence, all 30 steps, that is to be played when the board is swept from left to right. A `for` loop is used for accessing each element of the array, in order. A `while` loop is used to wait on Timer1 for the proper sequence timing.

1.1.2   Contains a pause for the sweep back, implemented using a `while` loop with a longer delay on Timer1.

Testing with the Logic Analyzer

To test the program, we will initially use the MPLAB SIM software simulator and the Logic Analyzer window.

1. Build the project (using the appropriate check list).

2. Open the Logic Analyzer window.

3. Click on the Channel button to add, in order, all the I/O pins from RA0 to RA7 connected to the row of LEDs.

The "MPLAB SIM Set-up" and "Logic Analyzer Set-up" checklists will help you make sure that you don't forget any detail.

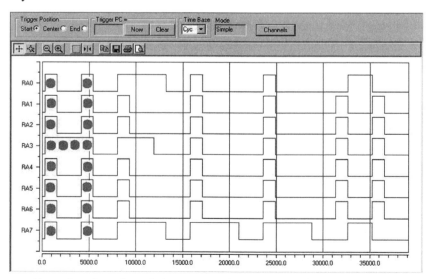

Figure 3-1. Snapshot of the Logic Analyzer window after the first sweep.

Then, I suggest you go back to the editor window and set the cursor on the first instruction of the 3.3.2 section and select the "Run to Cursor" option from the right click (context) menu. This will let the program execute the entire portion containing the message output (3.3.1) and stop before the long delay. As soon as the simulation halts on the cursor line, you can switch to the Logic Analyzer window and verify the output waveforms. They should look like the figure below:

To help you visualize the output, I added a few dots to represent the LEDs being turned on during the first few steps of the sequence. If you train your eye to see an LED on wherever the corresponding pin is at the logic high level, you should be able to read the desired message.

Using the Explorer16 demonstration board

If you have an actual Explorer16 demonstration board available, the fun can be doubled.

1. Use the "MPLAB ICD2 Set-up" checklist to enable the in-circuit debugger.
2. Use the "MPLAB ICD2 Device Configuration" to verify the device configuration bits proper setting for use on the Explorer16 demonstration board.
3. Use the "MPLAB ICD2 Programming" checklist to program the PIC24 in circuit.

If successful, and if you dim the light a bit in the room, you should be able to see the message flashing as you "shake" the board. The experience is going to be far from perfect though. With the Simulator and the Logic Analyzer window, we can choose which part of the sequence we want to visualize with precision and "freeze" it on the screen. On the demonstration board, you might find it quite challenging to synchronize the board's movement with the LED sequence.

Consider adjusting the timing constants to your optimal speed. After some experimentation, I found the values 100 and 800 ideal, respectively, for the short and long delays, but your preferences might differ.

Post-flight briefing

In this lesson we reviewed the declaration of a few basic variable types, including integers and floating point of different sizes. Array declarations and their initialization were also used to create an LED display sequence and a `for` loop was used to play it back.

Notes for assembly experts

If you were wondering whether the increment and decrement operators were going to be translated by the C30 compiler with the `inc` and `dec` assembly instructions, you were mostly right. I am saying "mostly" and not "always" because the ++ and -- operators are actually much smarter than that. If the variable they are applied to is an integer, as in the trivial examples above, this is certainly the case. But if they are applied to a pointer (which is a variable type that contains a memory address) they actually increase the address by the exact number of bytes required to represent the quantity pointed to. For example, a pointer to 16-bit integers will increment its address by 2, a pointer to a 32-bit long integer will increment its address by 4, and so on. To satisfy your curiosity, switch to the disassembly view and see how the MPLAB C30 chooses the best assembly code, depending on the situation.

Loops in C can be confusing: should you test the condition at the beginning or the end? Should you use the `for` type or not? The fact is, in some situations the algorithm you are coding will dictate which one to use, but in many situations you will have a degree of freedom and more than one type might do. Choose the one that makes your code more readable, and if it really doesn't matter, as in the main loop, just choose the one you like and be consistent.

Notes for PIC microcontroller experts

Depending on the target microcontroller architecture, and ultimately the arithmetic and logic unit (ALU), operating on bytes versus operating on word quantities can make a big difference in terms of code compactness and efficiency. While in the PIC16 and PIC18 8-bit architectures there is a strong incentive to use byte-sized integers wherever possible, in the PIC24, 16-bit architecture word-sized integers can be manipulated with the same efficiency. The only limiting factor preventing us from always using 16-bit integers with the MPLAB C30 compiler is the consideration of the relative preciousness of the internal resources of the microcontroller, and in this case the RAM memory.

Notes for C experts

Even if PIC24 microcontrollers have a relatively large RAM memory array, embedded-control applications will always have to contend with the reality of cost and size limitations. If you learned to program in C on a PC or a workstation, you probably never considered using anything smaller than an `int` as a loop index. Well, this is the time to think again. Shaving one byte at a time off the requirements of your application might, in some cases, mean the ability to select a smaller model of PIC24 microcontroller, saving fractions of a dollar that, when multiplied by thousands or millions of units (depending on your production run rates), can mean real money added to the bottom line. In other words, if you learn to keep the size of your variables to the strict minimum necessary, you will become a better embedded-control designer and ultimately…this is what engineering is all about.

Tips and tricks

This is the third lesson and, I am sure you will have noticed, for the third time I have been instructing you to start the simulation by setting a cursor on the first line of code and executing a Run To Cursor command (or setting a breakpoint) instead of more simply starting to single-step through the code. Why bother? Why can't we just start in Animation mode, for example, right after completing the project build?

As I briefly mentioned more than once, it is because of the C0 initialization code. Let me add, it's also because of MPLAB's obsessive desire to shield you from the low-level details. In fact, MPLAB won't even show the cursor (the big green arrow) if you try to single-step through it—quite a disconcerting experience. It will not let you see any trace of the C0 code even if you use the Disassembly window. But the C0 code is starting to do interesting things for you, and you might be getting curious. For example, in this last exercise we declared an array called `bitmap[]` and we asked for it to be initialized with a specific series of values. The array, being a data structure, resides in RAM during execution, so the compiler has to instruct the C0 initialization code to copy the contents of the array from a table in Flash memory immediately after the program start.

The only way to take a look at the C0 inner workings is to open the Program Memory window ("View→Program Memory"), select the Symbolic mode (using the buttons at the bottom of the window), and patiently inspect the assembly code. A few labels here and there will offer a little support. The first line of the program memory window will correspond to the reset vector of the PIC24 and will always contain a jump to the proper beginning of the program.

```
0000      goto _reset
```

You will have to scroll through several pages of what, you will learn shortly, is the interrupt vectors table. Eventually, you will find the _reset label. There, in a short sequence, you will recognize three essential pieces of code:

- the stack pointer (w15) initialization

  ```
  _reset mov.w #0x81e,w15
  ```

- a call to a subroutine for the variable (RAM) initialization

  ```
  rcall _data_init
  ```

- the call to the main() function

  ```
  call main
  ```

- a software reset instruction upon program termination

  ```
  reset
  ```

I hope this satisfies your curiosity for now. If during a future debugging session you are not able to find the cursor, chances are you will be able to find it in here. Something might have caused the processor to reset (a bug, an external event?) and you might be stepping through the very heart of the c0 initialization code. Check out the many emergency checklists created to help you recover and find your way safely home.

Exercises

1. Improve the display/hand synchronization, waiting for a button to be pressed before the hand sweep is started.

2. Add a switch to sense the sweep movement reversal and play the LED sequence backward on the back sweep.

Books

- Rony, P., Larsen D. and Titus J., 1976

  The 8080A Bugbook, Microcomputer Interfacing and Programming

  Howard W. Sams & Co., Inc., Indianapolis, IN

  This is the book that introduced me to the world of microprocessors and changed my life forever. No high-level language programming here, just the basics of assembly programming and hardware interfacing. (Too bad this book is already considered museum material; see link below.)

- Shulman, S. (2003)

  Unlocking the Sky, Glenn Hammond Curtis and the Race to Invent the Airplane

  Harper Collins, New York, NY

  A beautiful recount of the "struggle to innovate" in the early days of aviation.

Links

- *http://www.bugbookcomputermuseum.com/BugBook-Titles.html*

  A link to the "Bugbooks museum"—30 years since the introduction of the Intel 8080 microprocessor and it is like centuries have already passed.

NUMB3RS

The human sense of equilibrium is based on a device (the labyrinth or vestibular apparatus) located inside the ear that gives us feedback on gravity and motion. But, unlike that of birds, ours was just not designed for flight. It can be easily tricked by a little centrifugal acceleration and, in the absence of visual clues (in fog, clouds or during a night flight), it can have us flying happily into a tightening spiral…into the ground. To overcome our shortcomings, we have to rely on instruments to tell us how fast we are flying, in which direction and, perhaps most importantly, which way is up. Practically, this means that so much information that directly reaches the brain of a bird from its senses will arrive at our brain only in the form of numbers.

A good portion of the time spent by a student pilot on an airplane, after the first few flights, is spent learning the "right" numbers for his airplane—like the best climb speed, the best glide speed, the take-off (rotation) speed, the approach speeds and so on. Most of the time, these numbers are available inside the Pilot Operating Handbook (POH), the airplane datasheet, and, for convenience, on the related checklists. Each pilot tries his best to follow them religiously so that his flying performance gains consistency as he improves the command of the machine. However, even the most experienced aerobatic pilots, and certainly all the airline pilots who fly thousands of hours every year, will tell you how flying can be extremely spontaneous, if you know all your numbers exactly!

Similarly in embedded control, we need to know well the numeric types, their relative performance, and the costs and benefits of each one.

Flight plan

In this lesson we will review all the numerical data types offered by the MPLAB® C30 compiler. We will learn how much memory the compiler allocates for the numerical variables and we will investigate the relative efficiency of the routines used to perform arithmetic operations by using the MPLAB SIM Stopwatch as a basic profiling tool. This experience will help you choose the "right" numbers for your embedded-control application, understanding when and how to balance performance and memory resources, real-time constraints and complexity.

Preflight checklist

This entire lesson will be performed exclusively with software tools including the MPLAB IDE, MPLAB C30 compiler and the MPLAB SIM simulator.

Use the "New Project Set-up" checklist to create a new project called "Numbers" and a new source file called `numbers.c`.

The flight

To review all the data types available, I recommend you take a look at the MPLAB C30 User Guide. You can start in Chapter 5, where you can find a first list of the supported integer types.

Type	Bits	Min	Max
char, signed char	8	-128	127
unsigned char	8	0	255
short, signed short	16	-32768	32767
unsigned short	16	0	65535
int, signed int	16	-32768	32767
unsigned int	16	0	65535
long, signed long	32	-2^{31}	$2^{31} - 1$
unsigned long	32	0	$2^{32} - 1$
long long**, signed long long**	64	-2^{63}	$2^{63} - 1$
unsigned long long**	64	0	$2^{64} - 1$

** ANSI-89 extension

Table 4-1. Integer data types.

As you can see in Table 4-1, there are 10 different integer types as specified in the ANSI C standard including: char, int, short, long, and long long, both in the signed (default) and unsigned variant. The table shows the number of bits allocated specifically by the MPLAB C30 compiler for each type, and, for your convenience, spells out the minimum and maximum value that can be represented.

It is expected that, when the type is signed, one bit must be dedicated to the sign itself and the resulting numerical range is therefore halved. It is also interesting to note how the C30 compiler treats int and short as synonyms by allocating 16 bits for both of them. Both 8- and 16-bit quantities can be processed efficiently by the PIC24 arithmetic and logic unit (ALU), so that most of the arithmetic operations can be coded by the compiler using few and efficient instructions. The long integers are treated as 32-bit quantities, using four bytes, while the long long type (specified by the ANSI C extensions in 1989) requires eight bytes. Operations on long integers are performed by the compiler using short sequences of instructions inserted inline. So, there is a small performance penalty to pay for using long integers, and a proportionally larger penalty to pay for long long integers, that must be taken into account.

Let's see a first integer example; we'll start by typing the following code:

```
unsigned int i,j,k;

main ()
{
    i = 0x1234;    // assign an initial value to i
    j = 0x5678;    // assign an initial value to j
    k = i * j;     // perform the product and store the result in k
}
```

After building the project (Project→Build All or Ctrl + F10), we can open the Disassembly window ("View→Disassembly Listing") and take a look at the code generated by the compiler. Even without knowing the PIC24 instruction set in detail, we can recognize the two assignments. They are performed by loading the literal values to register w0 first and from there to the memory locations reserved for the variable i, and later for variable j.

```
                i = 1234;
204D20          mov.w #0x4d2,0x0000          // move literal value to W0
884290          mov.w 0x0000,0x0852          // move data from W0 to i

                j = 5678;
2162E0          mov.w #0x162e,0x0000         // move literal value to W0
8842A0          mov.w 0x0000,0x0854          // move data from W0 to j

                k = i * j;
804291          mov.w 0x0852,0x0002          // move data from i to W1
8042A0          mov.w 0x0854,0x0000          // move data from j to W0
B98800          mul.ss 0x0002,0x0000,0x0000
8842B0          mov.w 0x0000,0x0856          // move result   to k
```

The multiplication is performed by transferring the values from the locations reserved for the two integer variables i and j back to registers w0 and w1, and then performing a single mul instruction. The result, available in w0, is stored back into the locations reserved for k. Pretty straightforward.

On optimization (or lack thereof)

You will notice how the overall program, as compiled, is somewhat redundant. The value of j, for example, is still available in register w0 when it is reloaded again—just before the multiplication. Can't the compiler see that this operation is unnecessary?

In fact, the compiler does not see things this clearly—its role is to create "safe" code, avoiding (at least initially) any assumption and using standard sequences of instructions. Later on, if the proper optimization options are enabled, a second pass (or more) is performed to remove the redundant code. During the development and debugging phases of a project, though, it is always good practice to disable all optimizations as they might modify the structure of the code being analyzed and render single-stepping and breakpoint placement problematic. In the rest of this book, we will consistently avoid making use of any compiler optimization option; we will verify that the required levels of performance are obtained regardless. As a consequence, you will be able to execute all the examples presented in this and the following chapters using the C30 Compiler Student Edition, which is free and available on the companion CD-ROM.

Testing

To test the code, we can choose to work with the simulator from the Disassembly Listing window itself, single-stepping on each assembly instruction. Or we can choose to work from the C source in the editor window, single-stepping through each C language statement. In both cases, we can:

1. Set the cursor on the first line containing the initialization of the first variable, and perform a Run To Cursor command to let the program initialize and stop the execution just before the first instruction we want to observe.

2. Open the Watch window ("View→Watch") and select WREG0 in the SFR selection box, then click on the "Add SFR" button.

3. Repeat the operation for WREG1.

4. Select "i" in the symbol selection box, and click on the "Add Symbol" button.

5. Repeat the operation for j and k.

6. Use the "Step Over" function to execute the next few program lines, observing the effects on the registers and variables in the Watch window. As we noted before, when the value of a variable in the Watch window changes, it is conveniently highlighted in red.

If you need to repeat the test, perform a Reset ("Debugger→Reset→Processor Reset") and again place the cursor on the first line of code to analyze, followed by a new Run To Cursor command.

Going long

At this point, modifying only the first line of code, we can change the entire program to perform operations on long integer variables.

```
unsigned long i,j,k;

main ()
{
        i = 0x1234;    // assign an initial value to i
        j = 0x5678;    // assign an initial value to j
        k = i * j;     // perform the product and store the result in k

}
```

Rebuilding the project and switching again to the Disassembly Listing window (if you had the editor window maximized and you did not close the Disassembly Listing window, you could use the Ctrl + Tab command to quickly alternate between the editor and the Disassembly Listing), we can see how the newly generated code is considerably longer than the previous version. While the initializations are still straightforward, the multiplication is now performed using several more instructions.

```
               k = i * j;
8042C1         mov.w 0x0858,0x0002
8042E0         mov.w 0x085c,0x0000
B80A00         mul.uu 0x0002,0x0000,0x0008
8042C1         mov.w 0x0858,0x0002
8042F0         mov.w 0x085e,0x0000
B98800         mul.ss 0x0002,0x0000,0x0000
780105         mov.w 0x000a,0x0004
410100         add.w 0x0004,0x0000,0x0004
8042E1         mov.w 0x085c,0x0002
8042D0         mov.w 0x085a,0x0000
B98800         mul.ss 0x0002,0x0000,0x0000
410100         add.w 0x0004,0x0000,0x0004
780282         mov.w 0x0004,0x000a
884304         mov.w 0x0008,0x0860
884315         mov.w 0x000a,0x0862
```

The PIC24 arithmetic and logic unit can only process 16 bits at a time, so the 32-bit multiplication is actually performed as a sequence of 16-bit multiplications and additions. The sequence used by the compiler is generated with pretty much the same technique that we learned to use in elementary school, only performed on a word at a time rather than a digit at a time.

Note on the multiplication of long integers

In practice, to perform a 32-bit multiplication using 16-bit instructions, there should be four multiplications and two additions, but you will note how the compiler has actually inserted only three multiplication instructions. What is going on here?

The fact is that multiplying two long integers (32 bits each) will produce a 64-bit wide result. But in the example above, we have specified that the result will be stored in yet another long variable, therefore limiting the result to a maximum of 32 bits. Doing so, we have clearly left the door open for the possibility (not so remote) of an overflow, but we have also given the compiler the permission to ignore the most significant bits of the result. Knowing those bits are not going to be missed, the compiler has eliminated completely the fourth multiplication step—in a way, already optimizing the code.

Long long multiplication

Changing the variables declarations to the long long integer type (64-bit) is just as simple:

```
unsigned long long i,j,k;

main ()
{
    i = 0x1234;    // assign an initial value to i
    j = 0x5678;    // assign an initial value to j
    k = i * j;     // perform the product and store the result in k

}
```

Recompiling and inspecting the new code in the Disassembly Listing window reveals that this time the compiler has chosen a different approach. Instead of a longer sequence inserted inline, there are now only a few instructions to transfer the data into predefined registers and there is a call to a subroutine. The subroutine will appear in the disassembly listing, after all the main function code. This subroutine is clearly separated and identified by a comment line that indicates it is part of a library, a module called "muldi3.c". The source for this routine is actually available as part of the complete documentation of the C30 compiler and can be found in the subdirectory "src/libm/src/" under the same directory tree where the C30 compiler has been installed on your hard disk.

By selecting a subroutine in this case, the compiler has clearly made a compromise. Calling the subroutine means adding a few extra instructions and using extra space on the stack. On the other hand, fewer instructions will be added each time a new multiplication (among long long integers) is required in the program; therefore code space will be preserved.

Floating point

Beyond integer data types, the C30 compiler offers support for a few more data types that can capture fractional values—the floating-point data types. There are three types to choose from, corresponding to two levels of resolution: float, double and long double.

Notice how the MPLAB C30 compiler, by default, allocates the same number of bits for both the float and the double types, using the single precision floating-point format defined in the IEEE754 standard. Only the long double data type is treated as a true double-precision IEEE754 floating-point type.

Type	Bits	E Min	E Max	N Min	N Max
float	32	-126	127	2^{-126}	2^{128}
double*	32	-126	127	2^{-126}	2^{128}
long double	64	-1022	1023	2^{-1022}	2^{1024}

E = Exponent
N = Normalized (approximate)
* double is equivalent to long double if -fno-short-double is used.

Table 4-2. Floating points data types.

Notes for C experts

It is my belief that these floating-point settings were intentionally used by the MPLAB C30 designers to simplify and make more efficient the porting of complex math algorithms to embedded-control target applications. Most of the algorithms and libraries available in literature are designed for the performance and resources of personal computers and workstations, and make use of double-precision floating-point arithmetic whenever possible to maximize accuracy. Most often in embedded control, we are willing to compromise some of that accuracy for the level of performance necessary to achieve real-time response.

If needed, this behavior can be changed either locally, by turning doubles into long doubles in selected cases, or globally, by using special compiler options (open the "Project→Build Options→Project" dialog box, check the Use alternate Setting check box and add "-fno-short-double" to the edit box underneath).

Since the PIC24 doesn't have a hardware floating point unit (FPU), all operations on floating-point types must be coded by the compiler using floating-point arithmetic libraries whose size and complexity is considerably larger/higher than any of the integer libraries. You should expect a major performance penalty if you choose to use these data types, but, again, if the problem calls for fractional quantities to be taken into account, the C30 compiler certainly makes dealing with them easy.

Let's modify our previous example to use floating-point variables:

```
float i,j,k;

main ()
{
    i = 12.34;     // assign an initial value to i
    j = 56.78;     // assign an initial value to j
    k = i * j;     // perform the product and store the result in k

}
```

After recompiling and inspecting the Disassembly Listing window, you will notice that the compiler has immediately chosen to use a subroutine instead of inline code.

Changing the program again to use double-precision floating-point type, `long double`, produces very similar results. Only the initial assignments seem to be affected, and all we can see is a subroutine call.

The C compiler makes using any data type so easy that we might be tempted to always use the largest integer or floating-point type available, just to stay on the safe side and avoid the risk of overflows and underflows. On the contrary, choosing the right data type for each application can be critical in embedded control to balance performance and optimize the use of resources. In order to make an informed decision, we need to know more about the level of performance we can expect when choosing the various precision data types.

Measuring performance

Let's use what we have learned so far about simulation tools to measure the actual relative performance of the arithmetic libraries (integer and floating-point) used by the C30 compiler. We can start using the software simulator's (MPLAB SIM) built-in Stopwatch tool, with the following code:

```
//
//    Numbers
//

int    i1, i2, i3;
long   l1, l2, l3;
long long ll1, ll2, ll3;
float f1,f2, f3;
long double d1, d2, d3;

main ()
{
```

```
i1 = 1234;      // testing integers (16-bit)
i2 = 5678;
i3= i1 * i2;    // 1. int multiplication

l1 = 1234;      // testing long integers (32-bit)
l2 = 5678;
l3= l1 * l2;    // 2. long multiplication

ll1 = 1234;     // testing long long integers (64-bit)
ll2 = 5678;
ll3= ll1 * ll2;      // 3.

f1 = 12.34;     // testing single precision (32-bit) floating point
f2 = 56.78;
f3= f1 * f2;    // 4. single precision multiplication

d1 = 12.34;     // testing double precision (64-bit) floating point
d2 = 56.78;
d3= d1 * d2;    // 5. double precision multiplication
}
```

After compiling and linking the project, we can set the cursor on the line containing the first integer multiplication (`// 1.`) in the editor window and perform a Run To Cursor, to position the program counter for our test. Open the Stopwatch window ("Debugger→Stopwatch") and position the window according to your preferences (personally I like it docked to the bottom of the screen so that it does not overlap with the editor window and it is always visible and accessible).

Zero the Stopwatch timer and execute a Step-Over command ("Debug→StepOver", or press F8). As the Simulator completes updating the Stopwatch window, you can manually record the execution time required to perform the integer operation. The time is provided by the simulator in the form of a cycle count along with an indication in milliseconds derived by the cycle count multiplied by the simulated clock frequency, a parameter specified in the Debugger Settings ("Debugger→Settings→Osc/Trace" tab).

Proceed by setting the cursor over the next multiplication (`// 2.`), and execute a new Run To Cursor command or simply continue Stepping until you reach it. Again zero the Stopwatch, execute a Step-Over and record the second time. Continue until all five types have been tested.

Multiplication Test	Cycle Count	Performance relative to		
		int	long	Float
Integer	4	1	-	-
Long Integer	15	3.75	1	-
Long-Long Integer	99	24.75	6.6	-
Single Precision f.p.	121	30	8	1
Double Precision f.p.	317	79	21	2.6

Table 4-3. Relative Performance Test Results Using MPLAB C30 rev. 1.30
(all optimizations disabled).

In Table 4-3, I have recorded the results (cycle counts) in the first column and then added more columns to show the relative performance ratios, obtained by dividing the cycle count of each row by the cycle count recorded for the reference type. Don't be alarmed if you happen to record different values; several factors can affect the measure. Future versions of the compiler could possibly use more efficient libraries, and/or optimization features could be introduced or enabled at the time of testing.

Keep in mind that this type of test lacks any of the rigorousness required by a true performance benchmark. What we are looking for here is just a basic understanding of the impact on the performance we can expect from choosing to perform our calculations with one data type versus another. We are looking for the big picture—relative orders of magnitude. For that purpose, the table we just obtained can already give us some interesting indications.

As expected, 16-bit operations appear to be the fastest. Long-integer (32-bit) multiplications are about four times slower, while long-long-integer (64-bit) multiplications are one order of magnitude slower. Again, it was expected that single precision floating-point operations would require more effort than integer operations. Multiplying a 32-bit integer is only about four times slower than multiplying a 16-bit integer. However, multiplying 32-bit floating-point numbers is more than 30 times slower than multiplying 16-bit integers. That means it is eight times slower than the corresponding 32-bit integer multiplication, or about an order of magnitude. Going to double-precision floating-point (64-bit) though, only doubles the number of cycles. This tells us that, apparently, the double-precision floating-point libraries used by the compiler are more efficient than the corresponding 64-bit integer libraries.

So, when should we use floating point and when should we use integer arithmetic?

Beyond the obvious, from the little we have learned so far we can perhaps extract the following rules:

1. Use integers every time you can (i.e., when fractions are not required, or the algorithm can be rewritten for integer arithmetic).

2. Use the smallest integer type that will not produce an overflow or underflow.

3. If you have to use a floating-point type (fractions are required), expect an order-of-magnitude reduction in the performance of the compiled program.

4. Double-precision floating-point (`long double`) seems to only reduce the performance further by a factor of two.

Keep in mind also that floating-point types offer the largest value ranges, but also are always introducing approximations. As a consequence, floating-point types are not recommended for financial calculations. Use `long` or `long long` integers instead, and perform all operations in cents (instead of dollars and fractions).

Post-flight briefing

In this lesson, we have learned not only what data types are available and how much memory is allocated to them, but also how they affect the resulting compiled program—code size and the execution speed. We used the MPLAB SIM simulator Stopwatch function to measure the number of instruction cycles (and therefore time) required for the execution of a series of code segments. Some of the information gathered will, hopefully, be useful to guide our actions in the future when balancing our needs for precision and performance in embedded-control applications.

Notes for assembly experts

The brave few assembly experts that have attempted to deal with floating-point numbers in their applications tend to be extremely pleased and forever thankful for the great simplification achieved by the use of the C compiler. Single- or double-precision arithmetic becomes just as easy to code as integer arithmetic has always been.

When using integer numbers, though, there is sometimes a sense of loss of control, as the compiler hides the details of the implementation and some operations might become obscure or much less intuitive/readable. Here are some examples of conversion and byte-manipulation operations that can induce some anxiety:

1. Converting an integer type into a smaller/larger one.

2. Extracting or setting the most or least significant byte of a 16-bit data type.

3. Extracting or setting one bit out of an integer variable.

The C language offers convenient mechanisms for covering all such cases via implicit type conversions as in:

```
int         i;      // 16-bit
long        l;      // 32-bit
l = i;              // the value of i is transferred into the two LSB of l
                    // the two MSB of l are cleared
```

Explicit conversions might be required (called "type casting") in some cases where the compiler would otherwise assume an error, as in:

```
int         i;      // 16-bit
long        l;      // 32-bit
i = (int) l;        // (int) is a type cast that results in the two MSB of l
                    // to be discarded as l is treated as a 16-bit value
```

Bit fields are used to cover the conversion to and from integer types that are smaller than one byte. Bit fields are treated by the MPLAB C30 compiler with great efficiency and will result in the use of bit-manipulation instructions whenever possible. The PIC24 library files contain numerous examples of definitions of bit fields for the manipulation of all the control bits in the peripheral and the core special-function registers.

Here is an example extracted from the include file used in our project, where the Timer1 control register T1CON is defined and each individual control bit is exposed in a structure defined as T1CONbits:

```
extern unsigned int   T1CON;
extern  union {
  struct {
    unsigned :1;
    unsigned TCS:1;
    unsigned TSYNC:1;
    unsigned :1;
    unsigned TCKPS0:1;
    unsigned TCKPS1:1;
```

```
         unsigned TGATE:1;
         unsigned :6;
         unsigned TSIDL:1;
         unsigned :1;
         unsigned TON:1;
      };
      struct {
         unsigned :4;
         unsigned TCKPS:2;
      };
   } T1CONbits;
```

Notes for PIC microcontroller experts

The PIC microcontroller user, familiar with the 8-bit PIC microcontrollers and their respective compilers, will notice a considerable improvement in the performance, both with integer arithmetic and floating-point arithmetic. The 16-bit ALU available in the PIC24 architecture is clearly providing a great advantage by manipulating twice the number of bits per cycle, but the performance improvement is further accentuated by the availability of up to eight working registers, which make the coding of critical arithmetic routines and numerical algorithms more efficient.

Tips and tricks

Math libraries

The MPLAB C30 compiler supports several standard ANSI C libraries including:

- "limits.h", which contains many useful macros defining implementation-dependent limits, such as, for example, the number of bits composing a char type (CHAR_BIT) or the largest integer value (INT_MAX).

- "float.h", which contains similar implementation-dependent limits for floating-point data types, such as, for example, the largest exponent for a single-precision floating- point variable (FLT_MAX_EXP).

- "math.h", which contains trigonometric functions, rounding functions, logarithms and exponentials.

Complex data types

The MPLAB C30 compiler supports complex data types, as an extension of both integer and floating-point types. Here is an example declaration for a single-precision floating-point type:

```
__complex__ float z;
```

Notice the use of a double underscore before and after the keyword complex.

The variable z so defined has now a real and an imaginary part that can be individually addressed using the syntax: __real__ z and __imag__ z, respectively.

Similarly, the next declaration produces a complex variable of 16-bit integer type:

```
__complex__ int x;
```

Complex constants are easily created adding the suffix "i" or "j" as in the following examples:

```
x = 2 + 3j;
z = 2.0f + 3.0fj;
```

All standard arithmetic operations (+, −, *, /) are performed correctly on complex data types; additionally, the "~" operator produces the complex conjugate.

Complex types could be pretty handy in some types of applications, making the code more readable and helping avoid trivial errors. Unfortunately, as of this writing, the MPLAB IDE support of complex variables during debugging is only partial, giving access only to the "real" part through the Watch window and the mouse-over function.

Exercises

1. Write a program that uses Timer2 as a stopwatch for real-time performance measurements. If the width of Timer 2 is not sufficient:

    *   use the prescaler (and lose some of the lsb), or

    *   use Timer2 and Timer3 joined in the new 32-bit timer mode.

2. Test the relative performance of the division for the various data types.

3. Test the performance of the trigonometric functions relative to standard arithmetic operations.

4. Test the relative performance of the multiplication for complex data types.

Books

*   Gahlinger, P. M. (2000)

    The Cockpit, a Flight of Escape and Discovery

    Sagebrush Press, Salt Lake City, UT

    An interesting journey around the world, following the author in search of…his soul.

    Every instrument in the cockpit triggers a memory and starts a new chapter.

Links

*   *http://en.wikipedia.org/wiki/Taylor_series*

    If you are curious how the C compiler can approximate some of the functions in the math library.

Interrupts

In This Chapter

- Nesting of interrupts
- Traps
- A template and an example for Timer1 interrupt
- A real example with Timer1

- Testing the Timer1 interrupt
- The secondary oscillator
- The Real-time Clock Calendar (RTCC)
- Managing multiple interrupts

Every pilot is taught to keep his eyes constantly scanning the horizon, looking for visual clues about position and direction of flight and looking for other airplanes. But, he also needs to check the airplane instruments momentarily to verify his speed and altitude, and to keep an eye on the map. Now and then, there might be the need to focus longer on one of the inputs, and it is essential to learn how some instruments need a more frequent check than others, depending on the phase of flight and a number of other conditions. In other words, pilots need to learn to multitask, assigning the correct priority to each instrument and optimizing the use of time so as to stay always ahead of the machine.

For reasons of efficiency, size, and ultimately cost, in the embedded-control world the smallest applications, which happen to be implemented in the highest volumes, most often cannot afford the "luxury" of a multitasking operating system and use the interrupt mechanisms instead to "divide their attention" on the many tasks at hand.

Flight plan

In this lesson we will see how the MPLAB® C30 compiler allows us to easily manage the interrupt mechanisms offered by the PIC24 microcontroller architecture. After a brief review of some of the C language extensions and some practical considerations, we will present a short example of how to use the secondary (low-frequency) oscillator to maintain a real-time clock.

Preflight checklist

This entire lesson will be performed exclusively with software tools, including the MPLAB IDE, MPLAB C30 compiler and the MPLAB SIM simulator.

Use the "New Project Set-up" checklist to create a new project called "Interrupts" and a new source file similarly called "interrupts.c".

The flight

An interrupt is an internal or external event that requires quick attention from the CPU. The PIC24 architecture provides a rich interrupt system that can manage as many as 118 distinct sources of interrupts. Each interrupt source can have a unique piece of code, called the Interrupt Service Routine (ISR) directly associated via a pointer, also called a "vector," to provide the required response action. Interrupts can be completely asynchronous with the execution flow of the main program. They can be triggered at any point in time and in an unpredictable order. Responding quickly to interrupts is essential to allow prompt reaction to the trigger event and a fast return to the main program execution flow. Therefore, the goal is to minimize the interrupt latency, defined as the time between the triggering event and the execution of the first instruction of the Interrupt Service Routine (ISR). In the PIC24 architecture, the latency is not only very short but it is also fixed for each given interrupt source—only three instruction cycles for internal events and four instruction cycles for external events. This is a highly desirable quality that makes the PIC24 interrupt management superior to most other architectures.

The MPLAB C30 compiler helps manage the complexity of the interrupt system by providing a few language extensions. The PIC24 keeps all interrupt vectors in one large Interrupt Vector Table (IVT) and the MPLAB C30 compiler can automatically associate interrupt vectors with "special" user-defined C functions as long as a few limitations are kept in consideration, such as:

* They are not supposed to return any value (use type `void`).

* No parameter can be passed to the function (use parameter `void`).

* They cannot be called directly by other functions.

* Ideally, they should not call any other function.

The first three limitations should be obvious given the nature of the interrupt mechanism—since it is triggered by an external event, there cannot be parameters or a return value because there is no proper function call in the first place. The last limitation is more of a recommendation to keep in mind for efficiency considerations.

The following example illustrates the syntax that could be used to associate a function to the Timer1 interrupt vector:

```
void __attribute__ (( interrupt)) _T1Interrupt ( void)
{
  // interrupt service routine code here...

} // _InterruptVector
```

The function name `_T1Interrupt` was not an arbitrary choice, but is actually the predefined identifier for the Timer 1 interrupt as found in the Interrupt Vectors Table of the PIC24, (defined in the datasheet) and as coded in the linker script, the ".gld" file loaded for the current project.

The `__attribute__` `(( ))` mechanism is used by the C30 compiler in this and many other circumstances as a way to specify special features such as a C language extension. Personally, I find this syntax too lengthy and hard to read. I recommend the use of a couple of macros that can be found in each PIC24 include (".h") files and that greatly improve the code readability. In the following example, the `_ISR` macro is used to the same effect as the previous code snippet:

```
void _ISR _T1Interrupt (void)
{
  // interrupt service routine code here...

} // _InterruptVector
```

From Tables 5-1a and 5-1b, taken from the PIC24FJ128GA010 family datasheet, you can see which events can be used to trigger an interrupt. Among the external sources available for the PIC24FJ128GA010, there are:

- 5 × External pins with level trigger detection

- 22 × External pins connected to the Change Notification module

- 5 × Input Capture modules

- 5 × Output Compare modules

- 2 × Serial port interfaces (UARTs)

- 4 × Synchronous serial interfaces (SPI and I²C™)

- Parallel Master Port

Among the internal sources we count:

- 5 × 16-bit Timers

- 1 × Analog-to-Digital Converter

- 1 × Analog Comparators module

- 1 × Real-time Clock and Calendar

- 1 × CRC generator

Many of these sources in their turn can generate several different interrupts. For example, a serial-port interface peripheral (UART) can generate three type of interrupts:

- When new data has been received and is available in the receive buffer for processing.

- When data in the transmit buffer has been sent and the buffer is empty, ready and available, to transmit more.

- When an error condition has been generated and action might be required to re-establish communication.

Each interrupt source also has five associated control bits, allocated in various special-function registers (see Table 5-1):

- The Interrupt Enable bit (typically represented with a suffix -IE):

  – When cleared, the specific trigger event is prevented from generating interrupts.

  – When set, it allows the interrupt to be processed.

Interrupt Source	Vector Number	IVT Address	AIVT Address	Interrupt Bit Locations		
				Flag	Enable	Priority
ADC1 Conversion Done	13	00002Eh	00012Eh	IFS0<13>	IEC0<13>	IPC3<6:4>
Comparator Event	18	000038h	000138h	IFS1<2>	IEC1<2>	IPC4<10:8>
CRC Generator	67	00009Ah	00019Ah	IFS4<3>	IEC4<3>	IPC16<14:12>
External Interrupt 0	0	000014h	000114h	IFS0<0>	IEC0<0>	IPC0<2:0>
External Interrupt 1	20	00003Ch	00013Ch	IFS1<4>	IEC1<4>	IPC5<2:0>
External Interrupt 2	29	00004Eh	00014Eh	IFS1<13>	IEC1<13>	IPC7<6:4>
External Interrupt 3	53	00007Eh	00017Eh	IFS3<5>	IEC3<5>	IPC13<6:4>
External Interrupt 4	54	000080h	000180h	IFS3<6>	IEC3<6>	IPC13<10:8>
I2C1 Master Event	17	000036h	000136h	IFS1<1>	IEC1<1>	IPC4<6:4>
I2C1 Slave Event	16	000034h	000034h	IFS1<0>	IEC1<0>	IPC4<2:0>
I2C2 Master Event	50	000078h	000178h	IFS3<2>	IEC3<2>	IPC12<10:8>
I2C2 Slave Event	49	000076h	000176h	IFS3<1>	IEC3<1>	IPC12<6:4>
Input Capture 1	1	000016h	000116h	IFS0<1>	IEC0<1>	IPC0<6:4>
Input Capture 2	5	00001Eh	00011Eh	IFS0<5>	IEC0<5>	IPC1<6:4>
Input Capture 3	37	00005Eh	00015Eh	IFS2<5>	IEC2<5>	IPC9<6:4>
Input Capture 4	38	000060h	000160h	IFS2<6>	IEC2<6>	IPC9<10:8>
Input Capture 5	39	000062h	000162h	IFS2<7>	IEC2<7>	IPC9<14:12>
Input Change Notification	19	00003Ah	00013Ah	IFS1<3>	IEC1<3>	IPC4<14:12>
Output Compare 1	2	000018h	000118h	IFS0<2>	IEC0<2>	IPC0<10:8>
Output Compare 2	6	000020h	000120h	IFS0<6>	IEC0<6>	IPC1<10:8>
Output Compare 3	25	000046h	000146h	IFS1<9>	IEC1<9>	IPC6<6:4>
Output Compare 4	26	000048h	000148h	IFS1<10>	IEC1<10>	IPC6<10:8>
Output Compare 5	41	000066h	000166h	IFS2<9>	IEC2<9>	IPC10<6:4>
Parallel Master Port	45	00006Eh	00016Eh	IFS2<13>	IEC2<13>	IPC11<6:4>
Real-Time Clock/Calendar	62	000090h	000190h	IFS3<14>	IEC3<13>	IPC15<10:8>
SPI1 Error	9	000026h	000126h	IFS0<9>	IEC0<9>	IPC2<6:4>
SPI1 Event	10	000028h	000128h	IFS0<10>	IEC0<10>	IPC2<10:8>
SPI2 Error	32	000054h	000154h	IFS2<0>	IEC0<0>	IPC8<2:0>
SPI2 Event	33	000056h	000156h	IFS2<1>	IEC2<1>	IPC8<6:4>
Timer1	3	00001Ah	00011Ah	IFS0<3>	IEC0<3>	IPC0<14:12>
Timer2	7	000022h	000122h	IFS0<7>	IEC0<7>	IPC1<14:12>
Timer3	8	000024h	000124h	IFS0<8>	IEC0<8>	IPC2<2:0>
Timer4	27	00004Ah	00014Ah	IFS1<11>	IEC1<11>	IPC6<14:12>
Timer5	28	00004Ch	00014Ch	IFS1<12>	IEC1<12>	IPC7<2:0>
UART1 Error	65	000096h	000196h	IFS4<1>	IEC4<1>	IPC16<6:4>
UART1 Receiver	11	00002Ah	00012Ah	IFS0<11>	IEC0<11>	IPC2<14:12>
UART1 Transmitter	12	00002Ch	00012Ch	IFS0<12>	IEC0<12>	IPC3<2:0>
UART2 Error	66	000098h	000198h	IFS4<2>	IEC4<2>	IPC16<10:8>
UART2 Receiver	30	000050h	000150h	IFS1<14>	IEC1<14>	IPC7<10:8>
UART2 Transmitter	31	000052h	000152h	IFS1<15>	IEC1<15>	IPC7<14:12>

Table 5-1. Interrupt Vectors as implemented in the PIC24FJ128GA010 family.

At power on, all interrupt sources are disabled by default.

- The Interrupt Flag (typically represented with a suffix -IF). This single bit of data is set each time the specific trigger event is activated, independently by the status of the enable bit. Notice how, once set, it must be cleared (manually) by the user. In other words, it must be cleared before exiting the interrupt service routine, or the same interrupt service routine will be immediately called again.

- The priority level (typically represented with a suffix -IP). Interrupts can have up to 7 levels of priority. Should two interrupt events occur at the same time, the highest priority event will be served first. Three bits encode the priority level of each interrupt source. At any given point in time, the PIC24 execution priority level value is kept in the SR register in three bits referred to as IPL0..IPL2. Interrupts with a priority level lower than the current value of IPL will be ignored. At power on, all interrupt sources are assigned a default level of four and the processor priority is initially set at level zero.

Within an assigned priority level there is also a relative (default) priority among the various sources in the fixed order of appearance in the IVT table.

Nesting of interrupts

Interrupts can be nested, so that a lower-priority interrupt service routine can be interrupted by a higher-priority routine. This behavior can be controlled by the NSTDIS bit in the INTCON1 register of the PIC24.

When the NSTDIS bit is set, as soon as an interrupt is received the priority level of the processor (IPL) is set to the highest level (7) independently of the specific interrupt level assigned to the event. This prevents new interrupts from being serviced until the present one is completed. In other words, when the NSTDIS bit is set, the priority level of each interrupt is used only to resolve conflicts, should multiple interrupts occur simultaneously, and all interrupts are serviced sequentially.

Traps

Eight additional vectors occupy the first locations on top of the IVT table. They are used to capture special error conditions such as a failure of the selected CPU oscillator, an incorrect address (word access to odd address), stack underflow, or a divide by zero (math error).

Vector Number	IVT Address	Trap Source
0	000004h	Reserved
1	000006h	Oscillator Failure
2	000008h	Address Error
3	00000Ah	Stack Error
4	00000Ch	Math Error
5	00000Eh	Reserved
6	000010h	Reserved
7	000012h	Reserved

Table 5-2. TRAP vector details.

Since these types of errors have generally fatal consequences for a running application, they have been assigned fixed priority levels above the seven basic levels available to all other interrupts. This also means that they cannot be inadvertently masked (or delayed by the NSTDIS mechanism) and it provides an extra level of security for the application. The MPLAB C30 compiler associates all trap vectors with a single default routine that will produce a processor reset. You can change such behavior using the same technique illustrated for all generic interrupt service routines.

A template and an example for Timer1 interrupt

This all might seem extremely complicated, but we will quickly see that, by following some simple guidelines, we can put it to use in no time. Let's create a template, which we will reuse in future practical examples that demonstrate the use of the Timer1 peripheral module as the interrupt source. We will start by writing the interrupt service routine function:

```
// 1. Timer1 interrupt service routine
void _ISR _T1Interrupt( void)
{

    // insert your code here
    // ...

    // remember to clear the interrupt flag before exit
    _T1IF = 0;


} //T1Interrupt
```

We used the _ISR macro just like before and made sure to declare the function type and parameters as void. Remembering to clear the interrupt flag (_T1IF) before exiting the function is extremely important, as we have seen. In general, the application code should be very concise. The goal of any interrupt service routine is to perform a simple task quickly and efficiently in rapid response to an event. As a general rule, I would say that if you should find yourself writing more than a page of code (or contemplating calling other functions) you should most probably stop and reconsider the goals and the structure of your application. Lengthy calculations have a place in the main function and specifically in the main loop, not inside an interrupt service routine where time is at premium.

Let's complete the template with a few lines of code that we will add to the main function:

```
main()
{
    // 2. initializations
    _T1IP  = 4;          // set Timer1 priority, (4 is the default value)
    TMR1   = 0;          // clear the timer
    PR1    = period-1;   // set the period register

    // 2.1 configure Timer1 module clock source and sync setting
    T1CON  = 0x8000;     // check T1CON register options

    // 2.2 init the Timer1 Interrupt control bits
    _T1IF = 0;           // clear the interrupt flag, before
    _T1IE = 1;           // enable the T1 interrupt source

    // 2.3 init the processor priority level
    _IP = 0;             // 0 is the default value

    // 3. the main loop
    while( 1)
    {
```

```
        // your main code here...

    }       // main loop

} // main
```

In 2, we assign a priority level to the Timer1 interrupt source, although this might not be strictly necessary, as we know that all interrupt sources are assigned a default level-four priority at power on. We also clear the timer and assign a value to its period register.

In 2.1, we complete the configuration of the timer module, by turning the timer on with the chosen settings.

In 2.2, we clear the interrupt flag just before enabling the interrupt source.

The interrupt-trigger event for the timer module is defined as the instant the timer value reaches the value assigned to the period register. In that instant, the interrupt flag is set and the timer is reset to begin a new cycle. If the interrupt-enable bit is set as well, and the priority level is higher than the processor current priority (_IP), the interrupt service function is immediately called.

In 2.3, we initialize the processor priority level although, once more, this is not strictly necessary as the processor priority is initialized to zero by default at power on.

In 3.0, we will insert the main loop code. If everything goes as planned, the main loop will execute continuously, interrupted periodically by a brief call to the interrupt service routine.

A real example with Timer1

By adding only a couple of lines of code, we can turn this template into a more practical example where Timer1 is used to maintain a real-time clock, with tenths of a second, seconds and minutes. As a simple visual feedback we can use the lower 8 bits of PORTA as a binary display showing the seconds running. Here is what we need to add:

- Before 1., add the declaration of a few new integer variables that will act as the seconds and minutes counters:
  ```
  int dSec = 0;
  int Sec = 0;
  int Min = 0;
  ```

- In 1.2, have the interrupt service routine increment the counter:
  ```
  dSec++;
  ```
  A few additional lines of code will be added to take care of the carry-over into seconds and minutes.

- In 2, set the period register for Timer1 to a value that (assuming a 32-MHz clock) will give us a tenth of a second period between interrupts.
  ```
  PR1 = 25000-1; // 25,000 * 64 * 1 cycle (62.5ns) = 0.1 s
  ```

- Set PORTA lsb as output:
  ```
  TRISA = 0xff00;
  ```

- In 2.1, set the Timer1 prescaler to 1:64 to help achieve the desired period.
  ```
  T1CON = 0x8020;
  ```

59

- In 3., add code inside the main loop to continuously refresh the content of PORTA (lsb) with the current value of the milliseconds counter.

```
PORTA = Sec;
```

The new project is ready to build:

```
#include <p24fj128ga010.h>

int dSec = 0;
int Sec = 0;
int Min = 0;

// 1. Timer1 interrupt service routine
void _ISR _T1Interrupt( void)
{
    // 1.1 your code here
    dSec++;                 // increment the tens of a second counter
    if ( dSec > 9)              // 10 tens in a second
    {
        dSec = 0;
        Sec++;                  // increment the minute counter

        if ( Sec > 59)// 60 seconds make a minute
        {
            Sec = 0;

    // 1.2 increment the minute counter
            Min++;

            if ( Min > 59)// 59 minutes in an hour
                Min = 0;
        } // minutes
    } // seconds

    // 1.3 clear the interrupt flag
    _T1IF = 0;

} //T1Interrupt

main()
{
    // 2. init Timer 1, T1ON, 1:1 prescaler, internal clock source
    _T1IP = 4;     // this is the default value anyway
    TMR1 = 0;            // clear the timer
    PR1 = 25000-1;// set the period register
    TRISA = 0xff00;      // set PORTA lsb as output
```

```
// 2.1 configure Timer1 module
T1CON = 0x8020;        // enabled, prescaler 1:64, internal clock

// 2.2 init the Timer 1 Interrupt, clear the flag, enable the source
_T1IF = 0;
_T1IE = 1;

// 2.3 init the processor priority level
_IP = 0;               // this is the default value anyway

// 3. main loop
while( 1)
{
        // your main code here
        PORTA = Sec;

} // main loop

} // main
```

Testing the Timer1 interrupt

1. Open the Watch window (dock it to your favorite spot).

2. Add the following variables:

   – dSec, select from the Symbol pulldown box, then click on Add

   – TMR1, select from the SFR pulldown box, then click on Add

   – SR, select from the SFR pulldown box, then click on Add

3. Open the Simulator Stopwatch window ("Debugger→StopWatch").

4. Set a breakpoint on the first instruction of the interrupt response routine after 1.1.

5. Set the cursor on the line and from the right click menu select: Set Breakpoint, or simply double click. By setting the breakpoint here, we will be able to observe whether the interrupt is actually being triggered.

6. Execute a Run ("Debugger→Run" or F9). The simulation should stop quickly, with the program counter cursor (the green arrow) pointing right at the breakpoint inside the interrupt service routine.

So we did stop inside the interrupt service routine! This means that the trigger event was activated; that is, the Timer1 reached a count of 24,999 (remember though that the Timer1 count starts with 0, therefore 25,000 counts have been performed) which, multiplied by the prescaler value, means that $25,000 \times 64$ or exactly 1.6 million cycles, have elapsed.

The Stopwatch window will confirm that the total number of cycles executed so far is, in fact, slightly higher than 1.6 million. The Stopwatch count includes the time required by the initialization part of our program too. At the PIC24's execution rate (16 million instructions per second or 62.5 ns per cycle) this all happened in a tenth of a second!

From the Watch window, we can now observe the current value of processor priority level (IP). Since we are inside an interrupt service routine that was configured to operate at level four, we should be able to verify that bits 3, 4 and 5 of the status register (SR) contain exactly this value. For convenience, the MPLAB IDE shows the completely decoded contents of the status register in a small box, as part of the status bar located at the bottom of the main window.

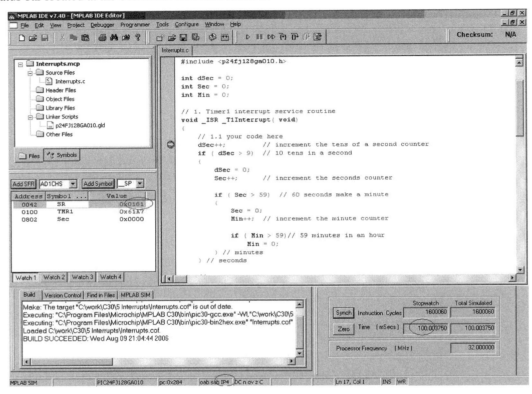

Figure 5-1. Screenshot showing the processor status after Timer1 interrupt.

In Figure 5-1, I have circled the IP indication in the status bar (showing IP4 to indicate interrupt priority level four) as well as the SR register content and the Stopwatch actual value (in milliseconds). Single stepping from the current position (using either the StepOver or the StepIn commands), we can monitor the execution of the next few instructions inside the interrupt service routine. Upon its completion, we can observe how the priority level returns back to the initial value—look for the IP0 indication in the status bar and the SR register bits 5, 6 and 7 to be cleared.

7. After executing another Run command, we should find ourselves again with the program counter (represented graphically by the green arrow) pointing inside the interrupt service routine. This time, you will notice how exactly 1.6 million cycles have been added to the previous count.

8. Add the Sec and Min variables to the Watch window.

9. Execute the Run command a few more times to verify that, after 10 iterations, the seconds counter is incremented.

To test the minutes increment, you might want to remove the current breakpoint and place a new one a few lines below—otherwise you will have to execute the Run command exactly 600 times!

10. Place the new breakpoint on the `Min++` statement in 1.2.

11. Execute Run once and observe that the seconds counter has already been cleared.

12. Execute the StepOver command once and the minute counter will be incremented.

The interrupt routine has been executed 600 times, in total, at precise intervals of one tenth of a second. Meanwhile, the code present in the main loop has been executed continuously to use the vast majority of the grand total of 960 million cycles. In all honesty, our demo program did not make much use of all those cycles—wasting them all in a continuous update of the PORTA content. In a real application, we could have performed a lot of work, all the while maintaining a precise real-time clock count.

The secondary oscillator

There is another feature of the PIC24 Timer1 module (common to all previous generations of 8-bit PIC® microcontrollers) that we could have used to obtain a real-time clock. In fact, there is a low-frequency oscillator (known as the secondary oscillator) that can be used to feed just the Timer1 module instead of the high-frequency main clock. Since it is designed for low-frequency operation (typically it is used in conjunction with an inexpensive 32,768-Hz crystal), it requires very little power to operate. And since it is independent from the main clock circuit, it can be maintained in operation when the main clock is disabled and the processor enters one of the many possible low-power modes. In fact, the secondary oscillator is an essential part for many of those low-power modes. In some cases it is used to replace the main clock, while in others it remains active only to feed the Timer1 or a selected group of peripherals.

To convert our previous example for use with the secondary oscillator, we will need to perform only a few minor modifications, such as:

- change the interrupt routine to count only seconds and minutes (the much slower clock rate does not require the extra step for the tenth of a second).

```
// 1. Timer1 interrupt service routine
void _ISR _T1Interrupt( void)
{
    // 1.1 clear the interrupt flag
    _T1IF = 0;

    // 1.2 your code here
    Sec++;          // increment the seconds counter

    if ( Sec > 59)// 60 seconds make a minute
    {
        Sec = 0;
        Min++; // increment the minute counter
```

```
        if ( Min > 59)// 59 minutes in an hour
                Min = 0;
    } // minutes
} //T1Interrupt
```

- in 2, change the period register to generate one interrupt every 32,768 cycles.

```
PR1 = 32768-1;// set the period register
```

- in 2.1, change the Timer1 configuration word (the prescaler is not required anymore).

```
T1CON = 0x8002;        // enabled, prescaler 1:1, use secondary oscillator
```

Unfortunately, you will not be able to immediately test this new configuration with the simulator, since the secondary oscillator input is not automatically simulated.

In a later lesson, we will learn how a new set of tools will help us to generate a stimulus file that could be used to provide a convenient emulation of a 32-kHz crystal connected to the T1CK and SOSCI pins of the PIC24.

The real-time clock calendar (RTCC)

Building on the previous two examples, we could evolve the real-time clock implementations to include the complete functionality of a calendar, adding the count of days, day of the week, months and years. These few new lines of code would be executed only once a day, once a month or once a year, and therefore would produce no decrease in the performance of the overall application whatsoever. Although it would be somewhat entertaining to develop such code, considering lapsed years and working out all the details, the PIC24FJ128GA010 already has a complete Real-time Clock and Calendar (RTCC) module built in and ready for use. How convenient! Not only does it feed from the same low-power secondary oscillator, but it comes with all the bells and whistles, including a built-in Alarm function that can generate interrupts. In other words, once the module is initialized, it is possible to activate the RTCC alarm and wait for an interrupt to be generated, for example, on the exact month, day, hour, minute and second desired once a year (or if set on February 29th, even once every four years!).

This is what the interrupt service routine would look like:

```
// 1. RTCC interrupt service routine
void _ISR _RTCCInterrupt( void)
{
        // 1.1 clear the interrupt flag
        _RTCIF = 0;

        // 1.2 your code here, will be executed only once a year
        // that is once every 365 x 24 x 60 x 60 x 16,000,000 MCU cycles
// that is once every 504,576,000,000,000 MCU cycles


} // RTCCInterrupt
```

Managing multiple interrupts

It is typical of an embedded-control application to require several interrupt sources to be serviced. For example, a serial communication port might require periodic attention at the same time that a PWM module is active and requires periodic updates to control an analog output. Multiple timer modules might be used simultaneously to produce pulsed outputs, while multiple inputs could be sampled by the analog-to-digital converter and their values would need to be buffered. There is almost no limit to the number of things that can be done with 118 interrupt sources available. At the same time, there is no limit to the complexity of the bugs that can be generated, thanks to the same sophisticated mechanisms, if a little discipline and some common sense are not applied.

Here are some of the rules to keep in mind:

1. Keep it short and simple. Make sure the interrupt routines are the shortest/fastest possible, and under no circumstances should you attempt to perform any processing of the incoming data. Limit the activity to buffering, transferring and flagging.

2. Use the priority levels to determine which event deserves to be serviced first, in case two events are triggered simultaneously.

3. But consider very carefully whether you want to face the additional complexity and headaches that result from enabling the use of nested interrupt calls. After all, if the interrupt service routines are short and efficient, the extra latency introduced by waiting for the current interrupt to be completed before a new one is serviced is going to be extremely small. If you determine that you don't need it that bad, make sure the NSTDIS control bit is set to prevent nesting:

```
_NSTDIS = 1; // disable interrupt nesting (default)
```

Post-flight briefing

In this lesson, we have seen how an interrupt service routine can be simple to code, thanks to the language extensions built into the C30 compiler and the powerful interrupt-control mechanisms offered by the PIC24 architecture. Interrupts can be an extremely efficient tool in the hands of the embedded-control programmer, to manage multiple tasks while maintaining precious timing and resources constraints. At the same time, they can be an extremely powerful source of trouble. In the PIC24 reference manual and the MPLAB C30 User Guide, you will find more useful information than we could possibly cram into one single lesson. Finally, in this lesson we took the opportunity to learn more about the uses of Timer1 and the secondary oscillator, and we got a glimpse of the features of the new Real-Time Clock and Calendar (RTCC) module.

Notes for C experts

The interrupt vector table (IVT) is an essential part of the C0 code segment for the PIC24. Actually two copies of it are required to be present in the first 256 locations of the program memory—one is used during normal program execution and the second (or Alternate IVT) during debugging. These tables account for most of the size of the C0 code in all the examples we have been developing in these first five lessons. Subtract 256 words (or 768 bytes) from the file size of each example to obtain the "net" code size.

Notes for assembly experts

The _ISRFAST macro can be used to declare a function as an interrupt service routine, and to further specify that it will use an additional and convenient feature of the PIC24 architecture: a set of four shadow registers. By allowing the processor to automatically save the content of the first four working registers (W0-W3—i.e., the most frequently used ones) and most of the content of the SR register in special reserved locations, without requiring the use of the stack, the shadow registers provide the fastest possible interrupt response time. Naturally, since there is only one set of such registers, their use is limited to applications where only one interrupt will be served at any given time. This does not limit us to use only one interrupt in the entire application, but rather to use _ISRFAST only in applications that have all interrupts with the same priority level or, if multiple levels are in use, reserve the _ISRFAST options only for the interrupt service routines with the highest level of priority.

Notes for PIC microcontroller experts

Notice that on the PIC24 architecture there is no single control bit that disables all interrupts, but there is an instruction (DISI) that can disable interrupts for a limited number of cycles. If there are portions of code that require all interrupts to be temporarily disabled, you can use the following inline assembly command:

```
__asm__ volatile("disi #0x3FFF"); // disable temporarily all interrupts

// your code here
// ...

DISICNT = 0; // re-enable all interrupts
```

Tips and tricks

According to the PIC24 datasheet, to activate the secondary low-power oscillator you need to set the SOSCEN bit in the OSCCON register. But before you rush to type the code in the last example and try to execute it on a real target board, notice that the OSCCON register, containing vital controls for the MCU affecting the choice of the main active oscillator and its speed, is protected by a locking mechanism. As a safety measure, you will have to perform a special unlock sequence first or your command will be ignored. Here is an example, using inline assembly:

```
// OSCCON unlock sequence, setting SOSCEN
    asm volatile ("mov    #OSCCON,W1");
    asm volatile ("mov.b  #0x46, W2");
    asm volatile ("mov.b  #0x57, W3");
    asm volatile ("mov.b  #0x02, W0");    // SOSCEN =1
    asm volatile ("mov.b  W2, [W1]");
    asm volatile ("mov.b  W3, [W1]");
    asm volatile ("mov.b  W0, [W1]");
```

A similar combination lock mechanism has been put in place to protect the key RTCC register RCFG-CAL. A special bit must be set (RTCWREN) to allow writing to the register, but this bit requires its own special unlock sequence to be executed first. Here is an example using, once more, inline assembly code:

```
// RCFGCAL unlock sequence, setting RTCWREN
asm volatile("disi      #5");
asm volatile("mov #0x55, w7");
asm volatile("mov w7,_NVMKEY");
asm volatile("mov #0xAA, w8");
asm volatile("mov w8,_NVMKEY");
```

```
asm volatile("bset        _RCFGCAL, #13");     // RTCWREN =1;
asm volatile("nop");
asm volatile("nop");
```

After these two steps, which initialize the RTCC, setting the date and time is trivial:

```
_RTCEN = 0;            // disable the module

// example set 12/01/2006 WED 12:01:30
_RTCPTR = 3;           // start the loading sequence
RTCVAL = 0x2006;       // YEAR
RTCVAL = 0x1100;       // MONTH-1/DAY-1
RTCVAL = 0x0312;       // WEEKDAY/HOURS
RTCVAL = 0x0130;       // MINUTES/SECONDS

// optional calibration
//_CAL = 0x00;

// enable and lock
_RTCEN = 1;            // enable the module
_RTCWREN = 0;          // lock settings
```

Setting the alarm does not require any special unlock combination. Here is an example that will help you remember my birthday:

```
// disable alarm
_ALRMEN = 0;

// set the ALARM for a specific day of the year (my birthday)
_ALRMPTR = 2;          // start the sequence
ALRMVAL = 0x1124;      // MONTH-1/DAY-1
ALRMVAL = 0x0006;      // WEEKDAY/HOUR
ALRMVAL = 0x0000;      // MINUTES/SECONDS

// set the repeat counter
_ARPT = 0;             // once
_CHIME = 1;            // indefinitely

// set the alarm mask
_AMASK = 0b1001;       // once a year

_ALRMEN = 1;           // enable alarm
_RTCIF = 0;            // clear interrupt flag
_RTCIE = 1;            // enable interrupt
```

Exercises

Write interrupt-based routines for the following applications:

1. Serial port software emulation.

2. Remote-control radio receiver.

3. NTSC video output (Hint: in a few chapters, you'll find the solution).

Books

- Curtis, K. E. (2006)

  Embedded Multitasking

  Newnes, Burlington, MA

  Keith knows multitasking and what it takes to create small and efficient embedded-control applications.

- Brown, G. (2003)

  Flying Carpet, The Soul of an Airplane

  Iowa State Press, Ames, IO

  Greg has many fun episodes from the real life of a general aviation pilot that uses his plane for recreation as well as family utility.

Links

- *http://www.aopa.org*

  This is the web site of the Aircraft Owners and Pilots Association. Feel free to browse through the web site and access the many magazines and free services offered by the association. You will find a lot of useful and interesting information in here.

Taking a look under the hood

In This Chapter

- ▶ *Memory space allocation*
- ▶ *Program space visibility*
- ▶ *Investigating memory allocation*

- ▶ *Looking at the MAP*
- ▶ *Pointers*
- ▶ *The heap*
- ▶ *MPLAB® C30 memory models*

Whether you are trying to get a driver's license or a pilot license, sooner or later you have to start looking under the hood, or the cowling for pilots. You don't have to understand how each part of the engine works, nor how to fix it—mechanics will be happy to do that for you. But a basic understanding of what is going on will help you be a better driver/pilot. If you understand the machine, you can control it better—it's that simple. You can diagnose little problems, and you can do a little maintenance.

Working with a compiler is not that dissimilar; sooner or later you have to start looking under the hood if you want to get the best performance out of it. Since the very first lesson, we have been peeking inside the engine compartment, but this time we will delve into a little bit more detail.

Flight plan

In this lesson we will review the basics of string declaration as an excuse to introduce the memory-allocation techniques used by the MPLAB C30 compiler. The RISC architecture of the PIC24 poses some interesting challenges and offers innovative solutions. We will use several tools, including the Disassembly Listing window, the Program Memory window and the Map file to investigate how the MPLAB C30 compiler and linker operate in combination to generate the most compact and efficient code.

Preflight checklist

This lesson will be performed exclusively with software tools including the MPLAB IDE, MPLAB C30 compiler and the MPLAB SIM simulator.

Use the "New Project Set-up" checklist to create a new project called "Strings" and a new source file similarly called "`strings.c`".

The flight

Strings are treated in C language as simple ASCII character arrays. Every character composing a string is assumed to be stored sequentially in memory in consecutive 8-bit elements of the array. After the last character of the string an additional byte containing a value of zero (represented in a character notation with '\0') is added as a termination flag.

> Notice though, that this is just a convention that applies to the standard C string manipulation library "string.h". It would be entirely possible, for example, to define a new library and store strings in arrays where the first element is used to record the length of the string—in fact, Pascal programmers would be very familiar with this method. Also, if you are developing "international" applications—i.e., applications that communicate using languages that require large character sets (like Chinese, Japanese, Korean)—you might want to consider using Unicode, a technology that allocates multiple bytes per character, in place of plain ASCII. The MPLAB C30 library "stdlib.h" provides basic support for the translation from/to multibyte strings according to the ANSI90 standard.

Let's get started by reviewing the declaration of a variable containing a single character:

```
char c;
```

As we have seen from the previous lessons, this is how we declare an 8-bit integer (character), that is treated as a signed value (−128. . . + 127) by default.

We can declare and initialize it with a numerical value:

```
char c = 0x41;
```

Or, we can declare and initialize it with an ASCII value:

```
char c = 'a';
```

Note the use of the single quotes for ASCII character constants. The result is the same, and to the C compiler there is absolutely no distinction between the two declarations—characters ARE numbers.

We can now declare and initialize a string as an array of 8-bit integers (characters):

```
char s[5] = { 'H', 'E', 'L', 'L', 'O'};
```

In this example, we initialized the array using the standard notation for numerical arrays. But, we could have also used a far more convenient notation (a shortcut) specifically created for string initializations:

```
char s[5] = "HELLO";
```

To further simplify things, and save you from having to count the number of characters composing the string (thus preventing human errors), you can use the following notation:

```
char s[] = "HELLO";
```

The MPLAB C30 compiler will automatically determine the number of characters required to store the string, while automatically adding a termination character (zero) that will be useful to the string manipulation routines later to correctly identify the string length. So, the example above is, in truth, equivalent to the following declaration:

```
char s[6] = { 'H', 'E', 'L', 'L', 'O', '\0' };
```

Assigning a value to a char (8-bit integer) variable and performing arithmetic on it is no different than performing the same operation on any integer type:

```
char c;// declare c as an 8-bit signed integer

c = 'a';   // assign to it the value corresponding to 'a' in the ASCII table
c ++;      // increment it...  it will represent the ASCII character 'b' now
```

The same operations can be performed on any element of an array of characters (string), but there is no simple shortcut, similar to the one used above for the initialization that can assign a new value to an entire string:

```
char s[15];        // declare s as a string of 15 characters

s = "Hello!"; // Error! This does not work!
```

Including the "string.h" file at the top of your source file, you'll gain access to numerous useful functions that will allow you to:

- copy the content of a string onto another:
  ```
  strcpy( s, "HELLO");        // s : "HELLO"
  ```
- append (or concatenate) two strings:
  ```
  strcat( s, " WORLD");// s : "HELLO WORLD"
  ```
- determine the length of a string:
  ```
  i = strlen( s);            // i : 11
  ```
- and many more.

Memory space allocation

Just as with numerical initializations, every time a string variable is declared and initialized as in:

```
char s[] = "Flying with the PIC24";
```

three things happen:

1. the MPLAB C30 linker reserves a contiguous set of memory locations in RAM (data space) to contain the variable: 22 bytes in the example above. This space is part of the ndata (near) data section.

2. the MPLAB C30 linker stores the initialization value in a 22-byte long table (in program memory). This space is part of the init code section.

3. the MPLAB C30 compiler creates a small routine that will be called before the main program (part of the C0 code we mentioned in previous chapters) to copy the values from code to data memory, therefore initializing the variable.

In other words, the string "Flying with the PIC24" ends up using twice the space you would expect, as a copy of it is stored in Flash program memory and space is reserved for it in RAM memory, too. Additionally, you must consider the initialization code and the time spent in the actual copying process. If the string is not supposed to be manipulated during the program, but is only used "as is," transmitted to a serial port or sent to a display, then there is no need to waste precious resources. Declaring the string as a "constant" will save RAM space and initialization code/time:

```
const char s[] = "Flying with the PIC24";
```

Now, the MPLAB C30 linker will only allocate space in program memory, in the `const` code section, where the string will be accessible via the Program Space Visibility window—an advanced feature of the PIC24 architecture that we will review shortly.

The string will be treated by the compiler as a direct pointer into program memory and, as a consequence, there will be no need to waste RAM space.

In the previous examples of this lesson, we saw other strings implicitly defined as constants:

```
strcpy( s, "HELLO");
```

The string "HELLO" was implicitly defined as of `const char` type, and similarly assigned to the `const` section in program memory to be accessible via the Program Space Visibility window.

Note that, if the same constant string is used multiple times throughout the program, the MPLAB C30 compiler will automatically store only one copy in the `const` section to optimize memory use, even if all optimization features of the compiler have been turned off.

Program space visibility

The PIC24 architecture is somewhat different from most other 16-bit microcontroller architectures you might be familiar with. It was designed for maximum efficiency according to the Harvard model, as opposed to the more common Von Neumann model. The big difference between the two is that there are two completely separated and independent buses available, one for access to the Program Memory (Flash) and one for access to the Data Memory (RAM). The net result is a doubled bandwidth; while the data bus is in use during the execution of one instruction, the program memory bus is available to fetch the next instruction code and initiate the decoding. In traditional Von Neumann architectures, the two activities must be interleaved instead, with a consequent performance penalty. The drawback of this architectural choice is that access to constants and data stored in program memory requires special considerations.

The PIC24 architecture offers two methods to read data from program memory: using special table access instructions (`tblrd`) and through a second mechanism, called the Program Space Visibility or PSV. This is a window of up to 32K bytes of program memory accessible from the data memory bus. In other words the PSV is a bridge between the program memory bus and the data memory bus.

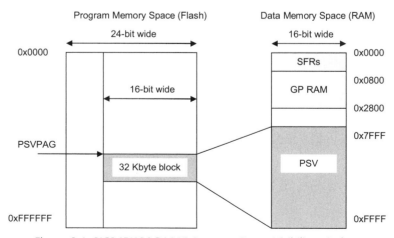

Figure 6-1. PIC24FJ128GA010 Program Space Visibility window.

Notice that the PIC24 uses a 24-bit wide program memory bus but operates only on a 16-bit wide data bus. The mismatch between the two buses makes the PSV "bridge" a little more interesting. In practice the PSV connects only the lower 16 bits of the program memory bus to the data memory bus. The upper portion (8 bits) of each program memory word is not accessible with the PSV window. On the contrary, when using the table-access instructions, all parts of the program memory word become accessible, but at the cost of having to differentiate the manipulation of data in RAM (using direct addressing) from the manipulation of data in program memory (using the special table-access instructions).

The PIC24 programmer can therefore choose between a more convenient but relatively memory-inefficient method for transferring data between the two buses such as the PSV, or a more memory-efficient but less-transparent solution offered by the table-access instructions.

The designers of the MPLAB C30 compiler considered the trade-offs and chose to use both mechanisms, although to solve different problems at different times:

- the PSV is used to manage constant arrays (numeric and strings) so that a single type of pointer (to the data memory bus) can be used uniformly for constants and variables.

- the table-access mechanism is used to perform the variable initializations (limited to the c0 segment) for maximum compactness and efficiency.

Investigating memory allocation

We will start investigating these issues with the MPLAB SIM simulator and the following short snippet of code:

```
/*
** Strings
*/

#include <p24fj128ga010.h>
#include <string.h>

// 1. variable declarations
const char a[] = "Learn to fly with the PIC24";
char b[100] = "";

// 2. main program
main()
{
    strcpy( b, "MPLAB C30");     // assign new content to b
} //main
```

Now, follow these steps:

1. Build the project using the Project Build checklist.

2. Add the Watch window (and dock it to the preferred position).

3. Select the two variables "a" and "b" from the symbol selection box and click "Add Symbol" to add them to the Watch window.

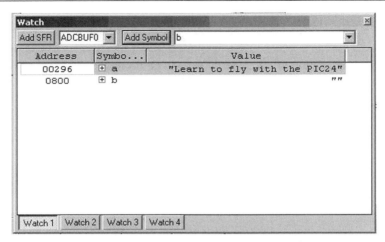

Figure 6-2. Adding arrays to the Watch window.

A little "+" symbol enclosed in a box will identify these variables as arrays and will allow you to expand the view to identify each individual element.

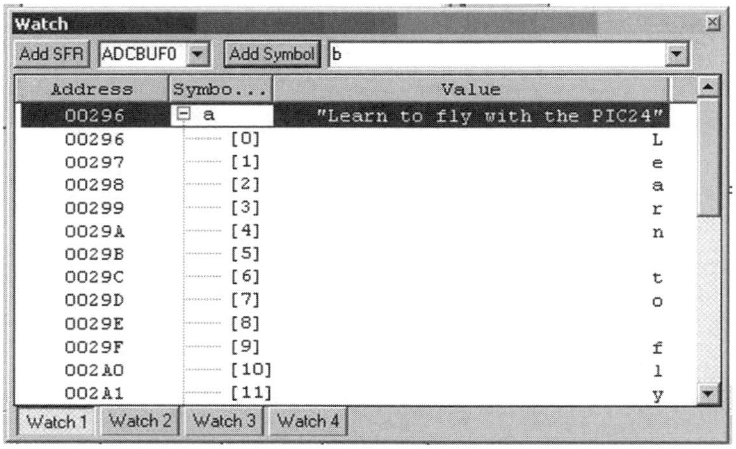

Figure 6-3. Expanding an array in the Watch window.

By default MPLAB shows each element of the array as an ASCII character, but you can change the display to reflect your personal preferences:

4. Select one element of the array with the left button of your mouse.

5. Right click to show the Watch window menu.

6. Select "Properties" (the last item in the menu).

You will be presented with the Watch window Properties dialog box.

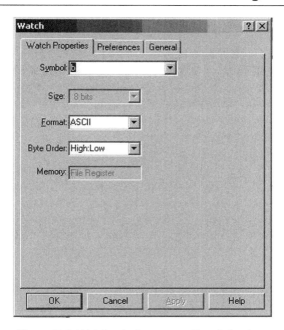

Figure 6-4. Watch window properties dialog box.

From this dialog box you can change the format used to display the content of the selected array element, but you can also observe the "Memory" field (grayed) that tells you where the selected variable is allocated: data or code space.

If you select the Properties dialog box for the constant string "a", you will notice that the memory space is indicated as "Program", confirming that the constant string is using only the minimum amount of space required in the Flash program memory of the PIC24 and will be accessed through the PSV so that no RAM needs to be assigned to it.

On the contrary, the Properties dialog box will reveal how the string "b" is allocated in a File register, or in other words RAM memory.

Continuing our investigation, notice how the string "a" appears to be already initialized, as the Watch window shows it's ready to use, right after the project build.

The string "b", on the contrary, appears to be still empty, and uninitialized. Only when we set the cursor on the first line of code inside the main routine and we execute a Run To Cursor command, the string "b" is initialized with the proper value.

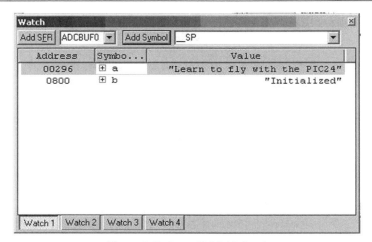

Figure 6-5. Array "b" initialized.

As we have seen, "b" is allocated in RAM space, and the C0 segment of code must be executed first for the variable to be initialized and "ready for use."

Just a warning—the Watch window aligns all strings to the right, so if there is a long string (like "a" in our example) and the window is too narrow, you might not be able to see the content of other variables containing shorter strings. Undock the Watch window if necessary, and resize it to be able to see the entire Value column.

Once more we can make use of the Disassembly Listing window to observe the code produced by the compiler:

```
---  C:\work\C30\6 Strings\Strings.c  ----------------------------------------
1:              /*
2:              ** Strings
3:              */
4:
5:              #include <p24fj128ga010.h>
6:              #include <string.h>
7:
8:              // 1. variable declarations
9:
10:             const char a[] = "Learn to fly with the PIC24";
11:             char b[100] = "Initialized";
12:
13:             // 2. main program
14:             main()
15:             {
 0028A  FA0000  lnk #0x0
16:                 strcpy( b, "MPLAB C30");    // assign new content to b
 0028C  282B21  mov.w #0x82b2,0x0002
 0028E  208000  mov.w #0x800,0x0000
```

```
00290   07FFF7      rcall 0x000280
17:
18:                 } // main
00292   FA8000      ulnk
00294   060000      return


---   c:\pic30-build\build_20060131\src\standardc\sxl\strcpy.c   ----------------

00280   780100      mov.w 0x0000,0x0004
00282   784931      mov.b [0x0002++],[0x0004]
00284   E00432      cp0.b [0x0004++]
00286   3AFFFD      bra nz, 0x000282
00288   060000      return
```

We can see the `main()` function and the `strcpy()` library function full disassembly appended at the bottom of the listing.

Notice how compact the code is produced for the `strcpy()` routine, barely five instructions. You will also appreciate how this is the only routine attached. Although the "`string.h`" library contains dozens of functions, and the include file "`string.h`" contains the declarations for all of them, the linker is wisely appending only the functions that are actually being used.

What the Disassembly Listing window does not show, though, is the initialization code `c0`. As mentioned in a previous chapter, in order to see it, you will have to rely on the Program Memory window (I recommend you select the Symbolic view tab at the bottom). There the most curious and patient readers will discover how the initialization of the string "b" is performed using the Table Read (`tblrd`) instructions to extract the data from the program memory (Flash) and to store the values in the allocated space in data memory (RAM).

Looking at the MAP

Another tool we have at our disposal to help us understand how strings (and in general any array variable) are initialized and allocated in memory is the "map file". This text file, produced by the MPLAB C30 linker, can be easily inspected with the MPLAB editor and is designed specifically to help you understand and resolve memory allocation issues.

To find this file, look for it in the main project directory where all the project source files are. Select "File→Open" and then browse until you reach the project directory. By default the MPLAB editor will list all the "`.c`" files, but you can change the File Type field to "`.map`".

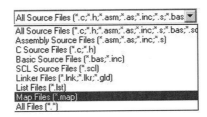

Figure 6-6. Selecting the ".map" file type.

77

Map files tend to be pretty long and verbose but, by learning to inspect only a few critical sections, you will be able to find a lot of useful data. The Program Memory Usage summary, for example, is found among the very first few lines:

```
Program Memory Usage

section        address     length (PC units)    length (bytes)  (dec)
-------        -------     -----------------    --------------  -----
.reset              0                   0x4                0x6   (6)
.ivt              0x4                  0xfc              0x17a  (378)
.aivt           0x104                  0xfc              0x17a  (378)
.text           0x200                  0x96               0xe1  (225)
.const          0x296                  0x26               0x39   (57)
.dinit          0x2bc                  0x4c               0x72  (114)
.isr            0x308                   0x2                0x3    (3)

            Total program memory used (bytes):           0x489 (1161) <1%
```

This is a list of small sections of code assembled by the MPLAB C30 linker in a specific order (dictated by the .gld linker script file) and position.

Most section names are pretty intuitive, other are...historical:

* the .reset section is where the reset vector is placed.

* the .ivt is the Interrupt Vector Table, discussed in the previous chapter.

* the .aivt is the Alternate Interrupt Vector Table.

* the .text section is where all the code generated by the MPLAB C30 compiler from your source files will be placed (the name of this section has been used since the original implementation of the very first C compiler).

* the .const section is where the constants (integers and strings) will be placed for access via the PSV.

* the .dinit section is where the variables initialization data (used by the C0 code) will be found.

* the .isr is where the Interrupt Service Routine (in this case a default one) will be found.

It's in the .const section that the "a" constant string, as well as the "MPLAB C30" (implicit) constant string, are stored for access via the PSV window.

You can confirm this by inspecting the Program Memory window at the address 0x296.

Note the two-by-two character grouping; remember how the PSV allows us to use only 16 bits of each 24-bit program memory word.

```
00290     ----    07FFF7  FA8000  060000  00654C  ........ ....Le..
00298     ----    007261  00206E  006F74  006620  ar..n .. to.. f..
002A0     ----    00796C  007720  007469  002068  ly.. w.. it..h ..
002A8     ----    006874  002065  004950  003243  th..e .. PI..C2..
002B0     ----    000034  00504D  00414C  002042  4...MP.. LA..B ..
002B8     ----    003343  000030  000800  000064  C3..0... ....d...
```

In .dinit is where the "b" variable initialization string, will be found. It is prepared for access via the table instructions, so it uses each and every one of the 24 bits available in each program memory word. Note the character grouping in three by three:

```
002C0    ----    000002 696E49 616974 7A696C  ....Ini. tia.liz.
002C8    ----    006465 000000 000000 000000  ed...... ........
002D0    ----    000000 000000 000000 000000  ........ ........
```

The next part of the map file we might want to inspect is the Data Memory Usage (RAM) summary:

```
Data Memory Usage

section         address       alignment gaps     total length  (dec)
-------         -------       --------------     ------------------
.ndata          0x800                      0              0x64  (100)


          Total data memory used (bytes):               0x64  (100) 1%
```

In our simple example, it contains only one section: .ndata, and in it, only one variable "b" for which 100 bytes are reserved starting at the address 0x800, the first location available in the PIC24 RAM.

Pointers

Pointers are variables used to refer indirectly (point to) other variables or part of their contents. Pointers and strings go hand in hand in C programming, as they are in general a powerful mechanism to work on any array data type. So powerful, in fact, that they are also one of the most dangerous tools in the programmer's hands and a source of a disproportionately large share of programming bugs. Some programming languages, like Java, have gone to the extreme of banning completely the use of pointers in an effort to make the language more robust and verifiable.

The MPLAB C30 compiler takes advantage of the PIC24 16-bit architecture to manage with ease large amounts of data memory (up to 32 kbytes of RAM available in current models). In particular, thanks to the PSV window, the MPLAB C30 compiler doesn't make any distinction between pointers to data memory objects and const objects allocated in program memory space. This allows a single set of standard functions to manipulate variables and/or generic memory blocks as needed from both spaces.

The following classic program example will compare the use of pointers versus indexing to perform sequential access to an array of integers:

```c
int *pi;        // define a pointer to an integer
int   i;        // index/counter
int a[10];      // the array of integers

// 1. sequential access using array indexing
for( i=0; i<10; i++)
    a[ i] = i;

// 2. sequential access using a pointer
pi = a;
for( i=0; i<10; i++)
{
```

```
        *pi = i;
        pi++;
    }
```

In 1. we performed a simple `for` loop and at each round in the loop we used "i" as an index in the array. To perform the assignment, the compiler will have to take the value of "i", multiply it by the size of the array element in bytes (2), and add the resulting offset to the initial address of the array "a" .

In 2. we initialized a pointer to point to the initial address of the array "a". At each round in the loop we simply used the pointer (*) to perform the assignment, and then we just incremented the pointer.

Comparing the two cases, we see how, by using the pointer, we can save at least one multiplication step for each round in the loop. If inside the loop the array element is used more times, the performance improvement is going to be proportionally greater.

Pointers syntax can become very "concise" in C, allowing for some effective code to be written, but also opening the door to more bugs.

At a minimum, you should become familiar with the most common contractions. The previous snippet of code is more often reduced to the following:

```
// 2. sequential access to array using pointers
for( i=0, p=a; i<10; i++)
    *pi++ = i;
```

Also note that an empty pointer—that is, a pointer without a target—is assigned a special value `NULL`, which is implementation specific and defined in "`stddef.h`".

The heap

One of the advantages offered by the use of pointers is the ability to manipulate objects that are defined dynamically (at run time) in memory. The "heap" is the area of data memory reserved for such use, and a set of functions, part of the standard C library "`stdlib.h`", provides the tools to allocate and free the memory blocks. They include at a minimum the fundamental functions:

```
void *malloc(size_t size);
```

takes a block of memory of requested size from the heap, and returns a pointer to it.

```
void free(void *ptr);
```

returns the block of memory pointed to by `ptr` to the heap.

The MPLAB C30 linker places the heap in the RAM memory space left unused above all project global variables and the reserved stack space. Although the amount of memory left unused is known to the linker and listed in the map file of each project, you will have to explicitly instruct the linker to reserve an exact amount for use by the heap.

Use the "Project→BuildOptions→Project" menu command to open the Build Options dialog box, select the MPLAB Link30 tab, and define the heap size in bytes.

As a general rule, allocate the largest amount of memory possible, as this will allow the `malloc()` function to make the most efficient use of the memory available. After all, if it is not assigned to the heap it will remain unused.

MPLAB C30 memory models

The PIC24 architecture allows for a very efficient (compact) instruction encoding for all operations performed on data memory within the first 8 kbytes of addressing space. This is referred to as the "near" memory area and in the case of the PIC24FJ128GA010 it corresponds to the group of SFRs (first 2 kbytes) and the following 6 kbytes of general-purpose RAM. Only the top 2 kbytes of RAM are actually outside the near space.

Access to memory beyond the 8-kbyte limit requires the use of indirect addressing methods (pointers) and could be less efficient if not properly planned. The stack (and with it all the local variables used by C functions) and the heap (used for dynamic memory allocation) are naturally accessed via pointers and are correspondingly ideal candidates to be placed in the upper RAM space. This is exactly what the linker will attempt to do by default. It will also try to place all the global variables defined in a project in the near memory space for maximum efficiency. If a variable cannot be placed within the near memory space, it has to be "manually" declared with a "far" attribute, so that the compiler will generate the appropriate access code. This behavior is referred to as the Small Data Memory Model as opposed to the Large Memory model, where each variable is assumed to be far unless the "near" attribute is specified.

In practice, while using the PIC24FJ128GA010, you will use almost uniquely the default Small Memory model and in rare occasions you will find it necessary to identify a variable with the "far" attribute. In lesson number 12, we will observe one such case, where a very large array that would otherwise not fit in the near memory space will have to be declared as "far". As a consequence, not only will the compiler generate the correct addressing instructions, but the linker will also push it to an upper area of RAM, giving priority to the other global variables and allowing them to be accessed in the near space.

Since access to elements of an array (explicitly via pointers or by indexing) is performed via indirect addressing anyway, there will be no performance or code size penalty.

A similar concept applies to the program memory space. In fact, within each compiled module, functions are called by making use of a more compact addressing scheme that relies on a maximum range of 32 kbytes. Program memory models (small and large) define the default behavior of the compiler/linker with regards to the addressing of functions within or outside such 32-kbyte range.

Post-flight briefing

In the C language, strings are defined as simple arrays of characters, but the C language standard had no concept of different memory regions (RAM vs. Flash) nor of the particular mechanisms required to cross the bridge between different buses in a Harvard architecture. The programmer using the MPLAB C30 compiler needs a basic understanding of the trade-offs of the various mechanisms available and the allocation strategies adopted to make the most of the precious resources (RAM especially) available to embedded-control applications.

Notes for C experts

The `const` attribute is normally used in the C language, together with most other variable types, only to assist the compiler in catching common parameter usage errors. When a parameter is passed to a function as a `const` or a variable is declared as a `const`, the compiler can in fact help flag any following attempt to modify it. The MPLAB C30 use of the PSV extends this semantic in a very natural way, allowing for a more efficient implementation, as we have seen.

Notes for assembly experts

The "string.h" library contains many useful block manipulation functions that can be useful, via the use of pointers, to perform operations on any type of data array, not just strings, like memcpy(), memcmp(), memset() and memmove().

The "ctype.h" library contains instead functions that help discriminate individual characters according to their position in the ASCII table, to discriminate lower case from upper case, and/or convert between the two.

Notes for PIC microcontroller experts

Since the PIC24 program memory is implemented using Flash technology, programmable with a single supply voltage even at run time, during code execution, it is possible to design boot-loaders—that is, applications that automatically update part or all of their own code. It is also possible to utilize a section of the Flash program memory as a nonvolatile memory storage area, within some pretty basic limitations. To write to the Flash program memory, though, you will need to utilize the table-access methods and exercise extreme caution. The PSV window is a read-only device and, as we have seen before, it gives access only to 16 of the 24 bits of each program memory location.

Also, notice that the memory can only be written in complete rows of 64 words each and must be first erased in blocks of 8 rows (512 words) each. This can make frequent updates impractical if single words or small data structures in general are being managed.

Tips and tricks

String manipulation can be fun in C once you realize how to make the zero termination character work for you efficiently. Take, for example, the mycpy() function below:

```
void mycpy( char *dest, char * src)
{
    while( *dest++ = *src++);
}
```

This is quite a dangerous piece of code, as there is no limit to how many characters could be copied, there is no check whether the dest pointer is pointing to a buffer that is large enough, and you can imagine what would happen should the src string be missing the termination character. It would be very easy for this code to continue beyond the allocated variable spaces and to corrupt the entire contents of the data RAM, including the all precious SFRs.

At a minimum, you should try to verify that pointers passed to your functions have been initialized before use. Compare them with the NULL value (declared in "stdlib.h" and/or "stddef.h") to catch the error.

Add a limit to the number of bytes to be copied; it is reasonable to assume that you will know the size of the strings/arrays used by your program, and if you don't, use the sizeof() operator. A better implementation of mycpy() could be the following:

```
void mycpy( char *dest, char *src, int max)
{
    if ((dest != NULL) && ( src != NULL))
        while (( max-- > 0) && ( *src))
            *dest++ = *src++;
}
```

Exercises

Develop new string manipulation functions to perform the following operations:

1. Search for a string in an array of strings, sequential.

2. Implement a Binary search.

3. Develop a simple Hash Table management library.

Books

*   Wirth, N. (1976)

    Algorithms + Data Structures = Programs

    Prentice-Hall, Englewood Cliffs, NJ

    With unparalleled simplicity, Wirth (the father of the Pascal programming language) takes you from the basics of programming all the way up to writing your own compiler.

Links

*   *http://en.wikipedia.org/wiki/Pointers#Support_in_various_programming_languages*

    Learn more about pointers and see how they are managed in various programming languages.

PART

II

Flying "Solo"

Part II – Flying "Solo"

Congratulations! You have endured the first few lessons and gained the necessary confidence to perform your first flight without an instructor sitting next to you. You are going to fly solo! As a consequence, in the next group of lessons more is going to be expected of you.

In the second part of this book, we will continue reviewing one by one the fundamental peripherals that allow a PIC24 to interface with the outside world. Since the complexity of the examples will grow a little bit, having an actual demonstration board at hand is recommended so that a practical demonstration can be performed. I will refer often to the standard Microchip Explorer16 demonstration board, but any third-party tool that offers similar features or allows for a small prototyping area can be used just as effectively.

Communication

On some of the major airlines, sometimes they make an additional channel available—the "cockpit channel," where you can listen to the actual conversation over the radio between the pilots and the traffic controllers. When you listen to it the first few times, it seems impossible to believe that there is actually any intelligent conversation going on. It all sounds like a continuous sequence of seemingly random numbers and unrecognizable acronyms. But, as you listen further and become familiar with some of the terms used in aviation, it starts to make sense. A precise protocol is followed by both pilots and controllers, selected radio frequencies are used as the media, and there is a whole language that must be learned and practiced to communicate from the cockpit of any airplane.

In embedded control, communication is equally a matter of understanding the protocols as well as the characteristics of the physical media available. In embedded-control programming, learning to choose the right communication interface can be as important as knowing how to use it.

Flight plan

In this lesson we will review a couple of communication peripherals available in all the general-purpose devices in the new PIC24 family. In particular, we will explore the asynchronous serial communication interfaces UART1 and UART2, and the synchronous serial communication interfaces SPI1 and SPI2, comparing their relative strengths and limitations for use in embedded-control applications.

Preflight checklist

In addition to the usual software tools, including the MPLAB® IDE, MPLAB C30 compiler and the MPLAB SIM simulator, this lesson will require the use of the Explorer16 demonstration board and the MPLAB ICD2 In Circuit Debugger.

Use the "New Project Set-up" checklist to create a new project called "SPI" and a new source file similarly called "spi2.c".

The flight

The PIC24FJ128GA010 offers seven communication peripherals that are designed to assist in all common embedded-control applications. As many as six of them are "serial" communication peripherals, as they transmit and receive a single bit of information at a time; they are:

- 2 × the universal asynchronous receiver and transmitters (UARTs)
- 2 × the SPI synchronous serial interfaces
- 2 × the I²C™ synchronous serial interfaces

The main difference between a synchronous interface (like the SPI or I²C) and an asynchronous one (like the UART) is in the way the timing information is passed from transmitter to receiver. Synchronous communication peripherals need a physical line (a wire) to be dedicated to the clock signal, providing synchronization between the two devices. The device(s) that originates the clock signal is typically referred to as the Master and the device(s) that synchronizes with it is called the Slave(s).

Synchronous serial interfaces

The I²C interface, for example, uses two wires (and therefore two pins of the microcontroller), one for the clock (referred to as SCL) and one (bidirectional) for the data (SDA).

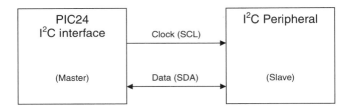

Figure 7-1. I²C interface block diagram.

The SPI interface instead separates the data line in two, one for the input (SDI) and one for the output (SDO), requiring one extra wire but allowing simultaneous (faster) data transfer in both directions.

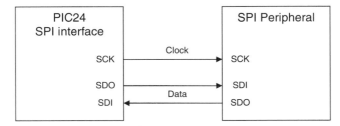

Figure 7-2. SPI interface block diagram.

In order to connect multiple devices to the same serial communication interfaces (bus configuration), the I²C interface requires a 10-bit address to be sent over the data line before any actual data is transferred. This slows down the communication but allows the same two wires (SCL and SDA) to be used for as many as (theoretically) 1,000 devices. Also, the I²C interface allows for multiple devices to act as masters and share the bus using a simple arbitration protocol.

The SPI interface, on the other side, requires an additional physical line, the slave select (SS) to be connected to each device. In practice, this means that, using an SPI bus, as the number of devices connected grows, the number of I/O pins required on the PIC24 grows proportionally with them.

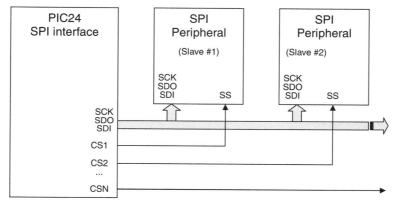

Figure 7-3. SPI bus block diagram.

Sharing an SPI bus among multiple masters is theoretically possible but practically very rare. The main advantages of the SPI interface are truly its simplicity and the speed that can be one order of magnitude higher than that of the fastest I²C bus (even without taking into consideration the details of the protocol-specific overhead).

Asynchronous serial interfaces

In asynchronous communication interfaces, there is no clock line, while typically two data lines are used: TX and RX, respectively, for input and output (optionally two more lines may be used to provide hardware handshake). The synchronization between transmitter and receiver is obtained by extracting timing information from the data stream itself. Start and stop bits are added to the data and precise formatting (with a fixed baud rate) allow reliable data transfer.

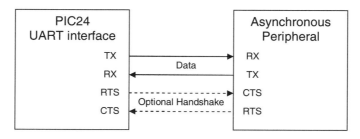

Figure 7-4. Asynchronous serial interface block diagram.

Several asynchronous serial interface standards dictate the use of special transceivers to improve the noise immunity, extending the physical distance up to several thousand feet.

Each serial communication interface has its advantages and disadvantages. Table 7-1 tries to summarize the most important ones as well as the most common applications:

	Synchronous		Asynchronous
Peripheral	SPI	I²C	UART
Max Bit Rate	10 Mbit/s	1 Mbit/s	500 kbit/s
Max Bus Size	Limited by no. of pins	128 devices	Point-to-point (RS232) 256 devices (RS485)
Number of Pins	3 + n × CS	2	2
Pros	Simple, low cost, high speed	Small pin count, allows multiple masters	Longer distance, improved noise immunity (requires transceivers)
Cons	Single master, short distance	Slowest, short distance	Requires accurate clock frequency
Typical Application	Direct connection to ASICs and other peripherals on same PCB	Bus connection with peripherals on same PCB	Interface with terminals, personal computers and other data acquisition systems
Examples	Serial EEPROMs (25CXXX series), MCP320X A/D converter, ENC28J60 Ethernet controller, MCP251X CAN controller...	Serial EEPROMs (24CXXX series), MCP98XX temperature sensors, MCP322x A/D converters...	RS232, RS422, RS485, LIN bus, MCP2550 IrDA interface...

Table 7-1. A comparison of synchronous and asynchronous serial communication peripherals.

Parallel interfaces

The Parallel Master Port (PMP) completes the list of basic communication interfaces of the PIC24. The PMP has the ability to transfer up to 8 bits of information at a time while providing several address lines, so as to interface directly to most LCD display modules commercially available (alphanumeric and graphic modules with integrated controller) as well as Compact Flash memory cards (or CF-I/O devices), printer ports and an almost infinite number of other basic 8-bit parallel devices available on the market and featuring the standard control signals: −CS, −RD, −WR.

In the rest of this lesson we will begin focusing specifically on the use of one synchronous serial interface, the SPI. In the following chapters we will cover also the asynchronous serial interfaces and separately the PMP.

Synchronous communication using the SPI modules

The SPI interface is perhaps the simplest of all the interfaces available, although the PIC24 implementation is particularly rich in options and interesting features.

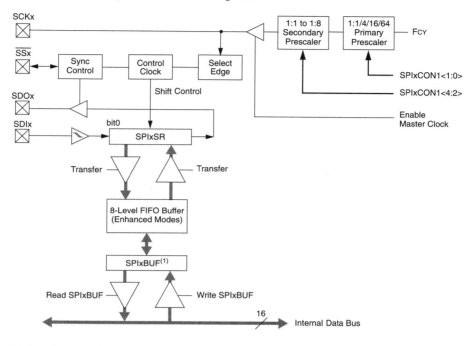

Note 1: In Standard modes, data is transferred directly between SPIxSR and SPIxBUF.

Figure 7-5. The SPI module block diagram.

The SPI interface is essentially composed of an 8-bit shift register: bits are simultaneously shifted in (MSB first) from the SDI line and shifted out from the SDO line in sync with the clock on pin SCK.

If the device is configured as a bus Master, the clock is generated internally (derived from the peripheral clock after a cascade of two prescalers for maximum flexibility) and output on the SCK pin. If the device is a bus Slave, the clock is received from the SCK pin.

As with all other peripherals we will encounter, the essential configuration options are controlled by a special function register, SPIxCON1 in this case, and additional advanced options are offered in SPIxCON2.

Upper Byte:

U-0	U-0	U-0	R/W-0	R/W-0	R/W-0	R/W-0	R/W-0
—	—	—	DISSCK	DISSDO	MODE16	SMP	CKE
bit 15							bit 8

Lower Byte:

R/W-0	R/W-0	R/W-0	R/W-0	R/W-0	R/W-0	R/W-0	R/W-0
SSEN	CKP	MSTEN	SPRE2	SPRE1	SPRE0	PPRE1	PPRE0
bit 7							bit 0

Figure 7-6. The SPIxCON1 control register.

To demonstrate the basic functionality of the SPI peripheral, we will use the Explorer16 demo board where the PIC24 SPI2 module is connected to a 25LC256 EEPROM device, often referred to as a Serial EEPROM (SEE or sometimes just E2—pronounced. e-squared). This is a small and inexpensive device that contains 256 kbits, or 32 kbytes, of nonvolatile high-endurance memory.

In order to prepare the SPI2 module for communication with the serial memory device, we will need to fine tune the peripheral module configuration.

The SEE responds to a short list of 8-bit (MOD16=0) commands that according to the device datasheet must be supplied via the SPI interface with the following setting:

- clock IDLE level is low, clock ACTIVE is high (CKP=0)

- serial output changes on transition from ACTIVE to IDLE (CKE=1)

The PIC24 will act as a bus Master (MSTEN=1) and will produce the clock signal SCK, deriving it from the internal clock after prescaling (in this case we will use the default prescalers values 1:64 and 1:8 for a total of 1:512).

The chosen configuration value can be defined as a constant that will later be assigned to the SPI2CON1 register:

```
#define SPI_MASTER  0x0120        // select 8-bit master mode, CKE=1, CKP=0
```

To enable the peripheral, we will access the SPI2STAT register where, similarly to most other PIC24 peripherals, bit 15 is the main enable control bit; another constant is defined for readability:

```
#define SPI_ENABLE 0x8000        // enable SPI port, clear status
```

Pin 12 of PORTD is connected to the memory chip select (CS), active low pin, so we will add two more definitions to the program, once more, to make it more readable:

```
#define CSEE     _RD12        // select line for Serial EEPROM
#define TCSEE    _TRISD12     // tris control for CSEE pin
```

We can now write the peripheral initialization part of our demonstration program:

```
// 1. init the PIC24 SPI peripheral
TCSEE = 0;              // make SSEE pin output
CSEE = 1;              // de-select the Serial EEPROM (low power standby)
SPI2CON1 = SPI_MASTER;  // select mode
SPI2STAT = SPI_ENABLE;  // enable the peripheral
```

We can now write a small function that will be used to transfer data to and from the serial EEPROM device:

```
// send one byte of data and receive one back at the same time
int writeSPI2( int data)
{
    SPI2BUF = data;                 // write to buffer for TX
    while( !SPI2STATbits.SPIRBF);   // wait for transfer to complete
    return SPI2BUF;                 // read the received value
}//writeSPI2
```

The `writeSPI2` is a truly bidirectional transfer function. It immediately writes a character to the transmit buffer and then enters a loop to wait for the receive flag to be set to indicate that the transmission was completed as well as that data was received back from the device. The data received is then returned as the value of the function.

When communicating with the memory device, however, there are situations when a command is sent to the memory, but there is no immediate response. There are also cases when data is read from the memory device but no further data commands need to be sent by the PIC24. In the first case (write command), the return value of the function can be simply ignored. In the second case (read command), a dummy value can be sent to the memory while shifting in data from the device.

The 25LC256 datasheet contains accurate depictions of all seven possible command sequences that can be used to read or write data to and from the memory device. A small table of constants can help encode all such commands:

```
// 25LC256 Serial EEPROM commands
#define SEE_WRSR   1                    // write status register
#define SEE_WRITE  2                    // write command
#define SEE_READ   3                    // read command
#define SEE_WDI    4                    // write disable
#define SEE_STAT   5                    // read status register
#define SEE_WEN    6                    // write enable
```

We can now write a small test program to verify that the communication with the device is properly established. For example, using the Read Status Register command we can interrogate the memory device and verify that the SPI peripheral is properly configured.

Testing the Read Status Register command

After sending the appropriate command (`SEE_STAT`), we will need to add an additional call to the `writeSPI2()` function with a dummy piece of data to capture the response from the memory device.

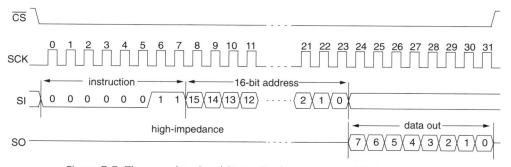

Figure 7-7. The complete Read Status Register command timing sequence.

Sending any command to the SEE requires at a minimum the following steps:

- Activate the memory, taking the CS pin low.
- Shift out the 8-bit instruction.

- Add one or more additional steps here, depending on the specific command.

- Deactivate the memory (taking the CS pin high) to complete the command, after which the memory will go back to a low-power consumption stand-by mode.

In practice, the following code is required to perform the complete Read Status Register operation:

```
// Check the Serial EEPROM status
CSEE = 0;                       // select the Serial EEPROM
writeSPI2( SEE_STAT);           // send a READ STATUS COMMAND, ignore immediate data
i = writeSPI2( 0);              // send dummy, read data
CSEE = 1;                       // deselect to complete the command
```

The complete project listing should look like:

```
/*
** SPI2 demo
*/
#include <p24fj128ga010.h>

// I/O definitions
#define CSEE    _RD12           // select line for Serial EEPROM
#define TCSEE   _TRISD12        // tris control for CSEE pin


// peripheral configurations
#define SPI_MASTER  0x0120      // select 8-bit master mode, CKE=1, CKP=0
#define SPI_ENABLE  0x8000      // enable SPI port, clear status

// 25LC256 Serial EEPROM commands
#define SEE_WRSR    1           // write status register
#define SEE_WRITE   2           // write command
#define SEE_READ    3           // read command
#define SEE_WDI     4           // write disable
#define SEE_STAT    5           // read status register
#define SEE_WEN     6           // write enable


// send one byte of data and receive one back at the same time
int writeSPI2( int data)
{
    SPI2BUF = data;                     // write to buffer for TX
    while( !SPI2STATbits.SPIRBF);       // wait for transfer to complete
    return SPI2BUF;                     // read the received value
}//writeSPI2
```

```
main()
{
        int i;

        // 1. init the SPI peripheral
        TCSEE = 0;                          // make SSEE pin output
        CSEE = 1;                           // de-select the Serial EEPROM
        SPI2CON1 = SPI_MASTER;              // select mode
        SPI2STAT = SPI_ENABLE;              // enable the peripheral

    // 2. Check the Serial EEPROM status
        CSEE = 0;                           // select the Serial EEPROM
        writeSPI2( SEE_STAT);               // send a READ STATUS COMMAND
        i = writeSPI2( 0);                  // send dummy, read data
        CSEE = 1;                           // terminate command <-set brkpt here

} // main
```

Follow the "MPLAB ICD2 Debugger Set-up" checklist to enable the In Circuit Debugger and prepare the project configuration. Then follow the "Project Build" checklist to compile and link the demo code.

1. After connecting the ICD2 to the Explorer16 demo board, program the PIC24 selecting the "Debugger→Program" option. By default MPLAB will choose the smallest range of memory required to transfer the project code into the device, so that programming time will be minimized. After a few seconds, the PIC24 should be programmed, verified and ready to execute.

2. Add the Watch window to the project.

3. Select "i" in the symbol selection box, and then click on the "Add Symbol" button.

4. Set the cursor on the last line of code in the main loop and set a breakpoint (double-click). Then start the execution by selecting the "Debugger→Run" command.

5. When the execution terminates, the contents of the 25LC256 memory Status Register should have been transferred to the variable "i", visible in the Watch window.

Unfortunately, you will be disappointed to learn that the default status of the 25LC256 memory (at power on) is represented by a 0x00 value, since BP1..BP0 are off to indicate no block protection, the write enable latch WEL is disabled, and no Write In Progress WIP flag should be active.

7	6	5	4	3	2	1	0
W/R	-	-	-	W/R	W/R	R	R
WPEN	x	x	x	BP1	BP0	WEL	WIP
W/R = writable/readable. R = read-only.							

Table 7-2. The 25LC256 Serial EEPROM Status Register.

Not a very telling result for our little test program. So, to spice things up a little we could start by setting the Write Enable Latch before interrogating the Status Register—it would be great to see bit 1 set.

To set the Write Enable Latch we will insert the following code before section 2, which we will promptly renumber to 2.2:

```
// 2.1 send a Write Enable command
CSEE = 0;               // select the Serial EEPROM
writeSPI2( SEE_WEN);    // send command, ignore immediate data
CSEE = 1;               // deselect to complete the command
```

1. Rebuild the project.

2. Reprogram the device.

3. Set a breakpoint to the last line of code in the main program, and

4. Run (or Run to Cursor).

If everything went well, you will see the variable "i" in the Watch window turn red and show a value of 2.

Now these are the great satisfactions that you can get only by developing code for a 16-bit embedded controller!

More seriously, now that the Write Enable latch has been set, we can add a write command and start "modifying" the contents of the EEPROM device. We can write a single byte at a time, or we can write a long string, up to a maximum of 64 bytes, all in a single sequence/command called Page Write. Read more on the datasheet about address restrictions that apply to this mode of operation, though.

Writing to the EEPROM

After sending the write command, two bytes of address (ADDR_MSB, ADDR_LSB) must be supplied before the actual data is shifted out. The following code exemplifies the correct write sequence:

```
// send a Write command
CSEE = 0;               // select the Serial EEPROM
writeSPI2( SEE_WRITE);  // send command, ignore immediate data
writeSPI2( ADDR_MSB);   // send MSB of memory address
writeSPI2( ADDR_LSB);   // send LSB of memory address
writeSPI2( data);       // send the actual data to be written
// send more data here to perform a page write
CSEE = 1;               // start actual EEPROM write cycle
```

Notice how the actual EEPROM write cycle initiates only after the CS line is brought high again. Also, it will be necessary to wait for a time (Twc), specified in the memory device datasheet, for the cycle to complete before a new command can be issued. There are two methods to make sure that the memory is allowed the right amount of time to complete the write command. The simplest one consists of inserting a fixed delay after the write sequence. The length of such a delay should be longer than the maximum cycle time specified in the memory device datasheet (Twc max = 5 ms).

A better method consists of checking the Status Register contents before issuing any further read/write command, waiting for the Write In Progress (WIP) flag to be cleared (this will also coincide with the Write Enable bit being reset). By doing so, we will be waiting only the exact minimum amount of time required by the memory device in the current operating conditions.

Reading the memory contents

Reading back the memory contents is even simpler; here is a snippet of code that will perform the necessary sequence:

```
// send a Write command
CSEE = 0;                    // select the Serial EEPROM
writeSPI2( SEE_READ);        // send command, ignore immediate data
writeSPI2( ADDR_MSB);        // send MSB of memory address
writeSPI2( ADDR_LSB);        // send LSB of memory address
data = writeSPI2( 0);        // send dummy, read data
// read more data here sequentially incrementing the address
CSEE = 1;                    // terminate the read sequence, return to low power
```

The read sequence can be indefinitely extended by reading sequentially the entire memory contents if necessary, and upon reaching the last memory address (0x7FFF), rolling over and starting from 0x0000 again.

A nonvolatile storage library

We can now assemble a small library of functions dedicated to access the 25LC256 serial EEPROM. The library will hide all the details of the implementation, such as the SPI port used, specific sequences and timing details. It will expose instead only two basic commands to read and write integer data types to a generic (black box) nonvolatile storage device.

Let's create a new project using the Project Wizard and the usual checklist. An appropriate name could be "NVM". After creating a new source file "nvm.c" we can copy most of the definitions we prepared in the SPI2 project:

```
/*
** NVM Access Library
*/

#include <p24fj128ga010.h>

#include "NVM.h"

// I/O definitions for PIC24 + Explorer16 demo board
#define CSEE      _RD12         // select line for Serial EEPROM
#define TCSEE     _TRISD12      // tris control for CSEE pin

// peripheral configurations
#define SPI_MASTER  0x0122      // select 8-bit master mode, CKE=1, CKP=0
#define SPI_ENABLE 0x8000       // enable SPI port, clear status

// 25LC256 Serial EEPROM commands
#define SEE_WRSR    1           // write status register
#define SEE_WRITE   2           // write command
#define SEE_READ    3           // read command
#define SEE_WDI     4           // write disable
#define SEE_STAT    5           // read status register
#define SEE_WEN     6           // write enable
```

From the same project you can extract: the initialization code, the SPI2 write function and the status register read command. Each one will become a separate function:

```
void InitNVM(void)
{
    // init the SPI peripheral
    TCSEE = 0;                      // make SSEE pin output
    CSEE = 1;                       // de-select the Serial EEPROM
    SPI2CON1 = SPI_MASTER;          // select mode
    SPI2STAT = SPI_ENABLE;          // enable the peripheral

}//InitNVM


int writeSPI2( int data)
{// send one byte of data and receive one back at the same time
    SPI2BUF = data;                 // write to buffer for TX
    while( !SPI2STATbits.SPIRBF);   // wait for transfer to complete
    return SPI2BUF;                 // read the received value
}//WriteSPI2


int ReadSR( void)
{// Check the Serial EEPROM status register
    int i;
    CSEE = 0;                       // select the Serial EEPROM
    WriteSPI2( SEE_STAT);           // send a READ STATUS COMMAND
    i = WriteSPI2( 0);              // send/receive
    CSEE = 1;                       // deselect to terminate command
    return i;
} //ReadSR
```

To create a function that reads an integer value from the nonvolatile memory, first we verify that any previous command (write) has been correctly terminated by reading the status register. A sequential read of two bytes is used to assemble an integer value:

```
int iReadNVM( int address)
{ // read a 16-bit value starting at an even address

    int lsb, msb;

    // wait until any work in progress is completed
    while ( ReadSR() & 0x3);    // check the two lsb WEN and WIP

    // perform a 16-bit read sequence (two byte sequential read)
    CSEE = 0;                   // select the Serial EEPROM
    WriteSPI2( SEE_READ);       // read command
    WriteSPI2( address>>8);     // address MSB first
```

```
        WriteSPI2( address & 0xfe);  // address LSB (word aligned)
        msb = WriteSPI2( 0);         // send dummy, read msb
        lsb = WriteSPI2( 0);         // send dummy, read lsb
        CSEE = 1;
        return ( (msb<<8)+ lsb);
}//iReadNVM
```

Finally, the write enable function can be created extracting the short segment of code used to access the Write Enable latch from our previous project and adding a page write sequence:

```
void WriteEnable( void)
{ // send a Write Enable command
    CSEE = 0;                            // select the Serial EEPROM
    WriteSPI2( SEE_WEN);                 // write enable command
    CSEE = 1;                            // deselect to complete the command
}//WriteEnable


void iWriteNVM( int address, int data)
{ // write a 16-bit value starting at an even address

    int lsb, msb;

    // wait until any work in progress is completed
    while ( ReadSR() & 0x3);             // check the two lsb WEN and WIP

    // Set the Write Enable Latch
    WriteEnable();

    // perform a 16-bit write sequence (2 byte page write)
    CSEE = 0;                            // select the Serial EEPROM
    WriteSPI2( SEE_WRITE);               // write command
    WriteSPI2( address>>8);              // address MSB first
    WriteSPI2( address & 0xfe);          // address LSB (word aligned)
    WriteSPI2( data >>8);                // send msb
    WriteSPI2( data & 0xff);             // send lsb
    CSEE = 1;
}//iWriteNVM
```

More functions could be added at this point to access long and long long types, for example, but for our purposes this will suffice.

Note that the "page write" operation (see the 25LC256 memory datasheet for details) requires the address to be aligned on a power of two boundary (in this case, just an even address will do). The requirement must be extended to the read function for consistency.

Save the code in file "nvm.c" file and add it to the project using one of the three methods shown in the checklists. You can either use the editor right-click menu and select "Add to Project" or right-click on the project window on the "Source Files" branch and choose "Add Files", and then select the "NVM.c" file from the current project directory.

To make a few selected functions from this module accessible to other applications, create a new file, "NVM.h", and insert the following declarations:

```
/*
** NVM storage library
**
** encapsulates 25LC256 Serial EEPROM
** as a NVM storage device for PIC24 + Explorer16 applications
*/

// initialize access to memory device
void InitNVM(void);

// 16-bit integer read and write functions
// NOTE: address must be an even value between 0x0000 and 0x7ffe
// (see page write restrictions on the device datasheet)
int   iReadNVM ( int address);
void iWriteNVM( int address, int data);
```

This will expose only the initialization function and the integer read/write functions, hiding all other details of the implementation.

Add the "NVM.h" file to the project by right clicking in the project windows on the Header Files icon and selecting it from the current project directory.

Testing the new NVM library

To test the functionality of the library we can now create a test application containing a few lines of code that repeatedly read the contents of a memory location (at address 0x1234), increment its value, and write it back to the memory:

```
/*
** NVM Library test
*/

#include <p24fj128ga010.h>

#include "NVM.h"


main()
{
    int data;

    // initialize the SPI2 port and CS to access the 25LC256
    InitNVM();
```

```
    // main loop
    while ( 1)
    {
        // read current content of memory location
        data = iReadNVM( 0x1234);

        // increment current value
        Nop();                     // <-set brkpt here
        data++;

        // write back the new value
        iWriteNVM( 0x1234, data);
        //address++;

    } // main loop
} //main
```

Save this file as "NVMtest.c" and add it to the current project too.

Invoking the Build All command, you will observe the MPLAB C30 compiler working sequentially on the two source files (.c) and later the linker to combine the object codes to produce an output executable (.hex).

We are planning on using the ICD2 as the debugging tool of choice to test this code, as the MPLAB SIM does not have the capability to accurately emulate the SPI ports. Make sure not only that it is selected in the Debugger menu, but also that in "Project→Settings", and specifically in the MPLAB C30 linker tab, the "Link for ICD2" option is selected. (See Figure 7-8.)

This setting is required when operating with the ICD2 debugger in order to reserve a few RAM locations (at the end of the memory space) for the ICD2 itself and avoid conflicts with the memory allocated by our application.

If the Build All command is completed successfully, the code is ready to be programmed on the device.

1. Adding data to the Watch window and setting a breakpoint on the line immediately following the read command will allow us to test the proper operation of the NVM library.

2. Hit the Run command and watch the program stop after the first read.

Note the value of data and then hit Run again. It should increment continuously and, even when resetting the program or disconnecting the board completely from the power supply to reconnect it later, we will observe that the contents of location 0x1234 will be preserved and successively incremented.

Careful—if the main program loop is left running indefinitely without any breakpoint, the test program will quickly turn into a test of the Serial EEPROM endurance. In fact the loop will continue to reprogram location 0x1234 at a rate that will be mostly dependent on the actual Twc of the device. In a best-case scenario (maximum Twc = 5 ms) this will mean 200 updates every second. Or, in other terms, the theoretical endurance limit of the EEPROM (1,000,000 cycles) will be reached in 5,000 seconds, or slightly less than one hour and a half of continuous operation.

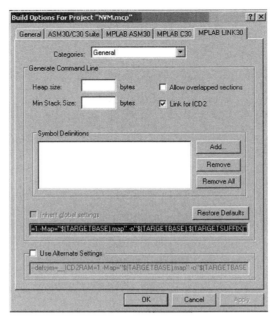

Figure 7-8. "Project→Build Options→MPLAB LINK30" Tab.

Post-flight briefing

In this lesson we have seen briefly how to use the SPI peripheral module, in its simplest configuration, to gain access to a 25LC256 Serial EEPROM memory, one of the most common types of nonvolatile memory peripherals used in embedded-control applications. The small library module developed will hopefully be useful to you in future applications to provide "mass" storage (32 kbytes).

Notes for C experts

The C programmer used to developing code for large workstations and personal computers will be tempted to develop the library further to include the most flexible and comprehensive set of functions. My word of advice is to resist, hold your breath and count to ten, especially before you start adding any new parameter to the library functions. In the embedded-control world, passing more parameters means using up more stack space, spending more time copying data to and from the stack and, in general, producing a larger output code. Keep the libraries simple and therefore easy to test and maintain. This does not mean that proper object-oriented programming practices should not be followed. On the contrary, the example above can be considered an example of object encapsulation, as all the details of the SPI interface and Serial EEPROM internal workings can be completely hidden from the user, who is provided with a simple interface to a generic storage device.

Notes for the experts

In developing the code examples above, we have ignored any access-speed consideration and simply configured the SPI module for the slowest possible operation. The PIC24 SPI peripheral module operates off the peripheral clock system, which can be ticking as fast as 16 MHz in the current production models. Few peripherals can operate at such speeds at 3V. Specifically, the 25LC256 series Serial

EEPROMs operate with a maximum clock rate of 5 MHz when the power supply is in the 2.5V to 4.5V range. This means that the fastest SPI port configuration compatible with the memory device can be obtained with a primary prescaler configured for a 4:1 ratio and a secondary prescaler configured for 1:1 operation (16 MHz/4 = 4 MHz). A sequential read command could therefore provide a maximum throughput of 4 Mbit per second or 512 kbytes per second. At such a rate the CPU would still be able to execute 32 instructions between each new byte of data received—not enough to perform complex calculations, but most probably sufficient for simple data transfer tasks.

Notes for PIC microcontroller experts

In addition to the SPI options available on most PIC microcontroller SPI interfaces (offered by the SSP and MSSP modules), such as:

* selectable clock polarity
* selectable clock edge
* master or slave mode operation

the PIC24 SPI interface module adds several new capabilities, including:

* 16-bit transfer mode
* data input sampling phase selection
* framed transmission mode
* frame sync pulse control (polarity and edge selectable)
* Enhanced Mode (8 deep transmit and receive FIFOs).

In particular, the 16-bit transfer mode could be used during sequential read and/or page write operations to improve the efficiency and increase the number of cycles available to the CPU between accesses to the SPI buffers (doubling it). But it is the Enhanced Mode, with eight-levels deep FIFOs, that can truly free up a considerable amount of CPU time. Up to eight words of data (16 bytes) can be written or retrieved from the SPI buffers in short bursts, leaving much more time to the CPU to process the data in between the successive bursts.

Tips and tricks

If you store important data in an external nonvolatile memory, you might want to put some additional safety measures in place (both hardware and software). From a hardware perspective make sure that:

* Adequate power-supply decoupling (a capacitor) is provided close to the memory device.
* A pull-up resistor (10k ohm) is provided on the Chip Select line, to avoid floating during the microcontroller power up and reset.
* An additional pull-down resistor (10k ohm) can be provided on the SCK clock line to avoid clocking of the peripheral during boundary scan and other board testing procedures.
* Verify clean and fast power-up and down slope are provided to the microcontroller to guarantee reliable power on reset operation. If necessary, add an external voltage supervisor (see MCP809 devices, for example).

A number of software methods can then be employed to prevent even the most remote possibility that a program bug or the proverbial cosmic ray might trigger the write routine. Here are some suggestions:

- Avoid reading and especially updating the NVM content right after power up. Allow a few milliseconds for the power supply to stabilize (application dependent).

- Add a software write-enable flag, and demand that the calling application set the flag before calling the write routine, possibly after verifying some essential application-specific entry condition.

- Add a stack-level counter; each function in the stack of calls implemented by the library should increment the counter upon entry and decrement it on exit. The write routine should refuse to perform if the counter is not at the expected level.

- Some users refuse to use the NVM memory locations corresponding to the first address (0x0000) and/or the last address (0xffff), believing they could be statistically more likely to be subject to corruption.

- More seriously, store two copies of each essential piece of data, performing two separate calls to the write routine. If each copy contains even a simple checksum, it will be easy, when reading it back, to discard the corrupted one and recover.

Exercises

1. Develop (circular) buffered versions of the read and write functions.

2. Enable the new SPI 16-bit mode to accelerate basic read and write operation.

3. Several functions in the library are performing locking loops that could reduce the overall application performance. By utilizing the SPI port interrupts implement a non blocking version of the library.

Books

*   Eady, F. (2004)

    Networking and Internetworking with Microcontrollers

    Newnes, Burlington, MA

    An entertaining introduction to serial communication in embedded control.

*   Buck, R. (1997)

    Flight of Passage: A Memoir

    Hyperion, New York, NY

    A grand adventure, in which two teenagers fly coast to coast in an aviation ritual of passage.

Links

*   *http://www.microchip.com/stellent/idcplg?IdcService=SS_GET_PAGE&nodeId=1406&dDocName=en010003*

    Use the link above or search on Microchip's web site for a free tool called "Total Endurance Software." It will help you estimate the endurance to expect from a given NVM device in your actual application conditions. It will give you an indication of the total number of e/w cycles or the number of expected years of your application life before a certain target failure rate is reached.

Asynchronous communication

If you have any experience with radio communication, whether it be with a walkie-talkie or a proper CB radio, you know how different it is from talking on a cell phone. For one, it's a half-duplex system, meaning you cannot talk if somebody else is already talking. You have to listen patiently, wait for your turn and then speak up, trying to be as concise as possible to give others the possibility of joining the conversation too. A simple verbal handshake system is used to prevent conflicts and misunderstanding.

This is exactly how it works in aviation. There is a precise protocol, a set of rules that dictates who should talk at any given point in time, what they should say and how. There are specific roles—such as the air traffic controllers, the pilots, the flight stations and the towers—and they all share the media in a coordinated and efficient way.

This works well as an introduction to the many asynchronous serial protocols. Some are full-duplex, other are just half-duplex, some are multipoint, others are point-to-point, but they all require coordination and adherence to basic rules (standards) that make communication possible and allow for efficient use of the media.

Flight plan

In this lesson we will review the PIC24 asynchronous serial communication interface modules, UART1 and UART2. We will develop a basic console library that will be handy in future projects for interface and debugging purposes.

Preflight checklist

In addition to the usual software tools, including the MPLAB® IDE, MPLAB C30 compiler and the MPLAB SIM simulator, this lesson will require the use of the Explorer16 demonstration board, the MPLAB ICD2 In-Circuit Debugger and a PC with an RS232 serial port (or a serial to USB adapter).

You will also need a terminal emulation program, if you are using Microsoft® Windows® operating system; the HyperTerminal application will suffice. ("Start→Programs→Accessories→Communication→HyperTerminal").

The flight

The UART interface is perhaps the oldest interface used in the embedded-control world. Some of its features were dictated by the need for compatibility with the first mechanical teletypewriters; this means that at least some of its technology has century-old roots.

On the other hand, nowadays finding an asynchronous serial port on a new computer (and especially on a laptop) is becoming a challenge. The serial port has been declared a "legacy interface" and, for several years now, strong pressure has been placed on computer manufacturers to replace it with the USB interface. Despite the decline in their popularity, and the clearly superior performance and characteristics of the USB interface, asynchronous serial interfaces are strenuously resisting in the world of embedded applications because of their great simplicity and extremely low cost of implementation.

Four main classes of asynchronous serial applications are still being used:

1. RS232 point-to-point connection: often simply referred to as "the serial port," is used by terminals, modems and personal computers, using +12V/–12V transceivers.

2. RS485 (EIA-485) multipoint serial connection: used in industrial applications, it uses a 9-bit word and special half-duplex transceivers.

3. LIN bus: a low-cost, low-voltage bus designed for noncritical automotive applications. It requires a UART capable of baud rate auto-detection.

4. Infrared wireless communication: requires a 38–40-kHz signal modulation and optical transceivers.

The PIC24's UART modules can support all four major application classes and packs a few more interesting features too.

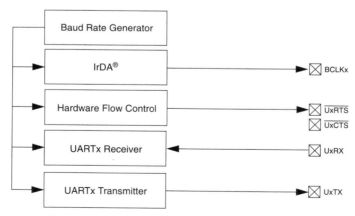

Figure 8-1. Simplified UART modules block diagram.

To demonstrate the basic functionality of a UART peripheral, we will use the Explorer16 demo board where the UART2 module is connected to an RS232 transceiver device and to a standard 9 poles D female connector. This can be connected to any PC serial port or, in absence of the "legacy interface" as mentioned above, to an RS232-to-USB converter device. In both cases, the Microsoft Windows HyperTerminal program will be able to exchange data with the Explorer16 board with a basic configuration setting.

The first step is the definition of the transmission parameters. The options include:

- baud rate
- number of data bits
- parity bit, if present
- number of stop bits
- handshake protocol.

For our demo we will choose the fast and convenient configuration: "115200, 8, N, 1, CTS/RTS", that is:

- 115,200 baud
- 8 data bits
- No parity
- 1 stop bit
- hardware handshake using the CTS and RTS lines.

UART configuration

Use the "New Project Set-up" checklist to create a new project called "Serial" and a new source file similarly called "serial.c". We will start by adding a few useful I/O definitions to help us control the hardware handshake lines:

```
/*
** Asynchronous Serial Communication
** UART2 RS232 asynchronous communication demonstration code
*/

#include <p24fj128ga010.h>

// I/O definitions for the Explorer16
#define CTS        _RF12       // Clear To Send, input, HW handshake
#define RTS        _RF13       // Request To Send, output, HW handshake
#define    TRTSTRISFbits.TRISF13   // Tris control for RTS pin
```

The hardware handshake is especially necessary when communicating with a Windows terminal application, since Windows is a multitasking operating system and its applications can sometimes experience long delays that would otherwise cause significant loss of data. We will use one I/O pin as

111

an input (RF12 on the Explorer16 board) to sense when the terminal is ready to receive a new character (Clear To Send), and one I/O pin as an output (RF13 on the Explorer16 board) to advise the terminal when our application is ready to receive a character (Request To Send).

To set the baud rate, we get to play with the Baud Rate Generator (BREG2), a 16-bit counter that feeds on the peripheral clock circuit. From the device datasheet, we learn that in the normal mode of operation (BREGH = 0) it operates off a 1:16 divider, versus a high-speed mode (BREGH = 1) where its clock operates off a 1:4 divider. A simple formula, published on the datasheet, allows us to calculate the ideal setting for our configuration:

```
BREG2 =  (Fosc / 8 / baudrate) -1      ; for BREGH=1
```

In our case this translates to the following expression:

```
BREG2 = (Fosc / 8 / 115,200) -1 = 33.7 where Fosc = 32MHz.
```

To decide how to best round out the result (we need a 16-bit integer after all) we will use the reverse formula to calculate the actual baud rate and determine the percentage error:

```
Error = ((Fosc/ 8 / (BREG2 + 1)) - baudrate) / baudrate %
```

Rounding up to a value of 34, we obtain an actual baud rate of 114,285 Bd with an error of just 0.7%, well within acceptable tolerance. With a value of 33, we obtain 117,647 baud or a 2.1% error, outside the acceptable tolerance range (± 2%) for a standard RS232 port.

We therefore define the constant BRATE as:

```
#define BRATE   34            // 115200 Bd (BREGH=1)
```

Two more constants will help us define the initialization values for the UART2 main control registers, called U2MODE and U2STA.

Upper Byte:							
R/W-0	U-0	R/W-0	R/W-0	R/W-0	U-0	R/W-0[1]	R/W-0[1]
UARTEN	—	USIDL	IREN	RTSMD	—	UEN1	UEN0
bit 15							bit 8

Lower Byte:							
R/W-0 HC	R/W-0	R/W-0 HC	R/W-0	R/W-0	R/W-0	R/W-0	R/W-0
WAKE	LPBACK	ABAUD	RXINV	BRGH	PDSEL1	PDSEL0	STSEL
bit 7							bit 0

Figure 8-2. The UxMODE control registers.

The initialization value for U2MODE will include the BREGH bit, the number of stop bits and the parity bit settings.

```
#define U_ENABLE 0x8008          // enable UART, BREGH=1, 1 stop, no parity
```

The initialization for U2STA will enable the transmitter and clear the error flags:

```
#define U_TX    0x0400           // enable transmission, clear all flags
```

Upper Byte:							
R/W-0	R/W-0	R/W-0	U-0	R/W-0 HC	R/W-0	R-0	R-1
UTXISEL1	UTXINV[1]	UTXISEL0	—	UTXBRK	UTXEN	UTXBF	TRMT
bit 15							bit 8

Lower Byte:							
R/W-0	R/W-0	R/W-0	R-1	R-0	R-0	R/C-0	R-0
URXISEL1	URXISEL0	ADDEN	RIDLE	PERR	FERR	OERR	URXDA
bit 7							bit 0

Figure 8-3. The UxSTA control registers.

We will create a new function, by using the constants defined above, to initialize the control register of the UART2, the baud-rate generator and the I/O pins used for the handshake:

```
void initU2( void)
{
    U2BRG  = BRATE;     // initialize the baud rate generator
    U2MODE = U_ENABLE;  // initialize the UART module
    U2STA  = U_TX;      // enable the Transmitter
    TRTS   = 0;         // make RTS an output pin
    RTS    = 1;         // set RTS default status (not ready)
} // initU2
```

Sending and receiving data

Sending a character to the serial port is a three-step procedure:

1. Make sure that the terminal (PC running Windows HyperTerminal) is ready. Check the Clear to Send (CTS) line. CTS is an active low signal—that is, while it is high, we better wait patiently.

2. Make sure that the UART is not still busy sending some previous data. PIC24 UARTs have a four-level deep FIFO buffer, so all we need to do is wait until at least the top level frees up, or in other words, we need to check for the transmit buffer full flag UTXBF to be clear.

3. Finally, transfer the new character to the UART transmit buffer (FIFO).

All of this can be nicely packaged in one short function:

```
int putU2( int c)
{
    while ( CTS);               // wait for !CTS, clear to send
    while ( U2STAbits.UTXBF);   // wait while Tx buffer full
    U2TXREG = c;
    return c;
} // putU2
```

To receive a character from the serial port, we follow a very similar sequence:

1. Alert the terminal that we are ready to receive by asserting the RTS signal (active low).

2. Patiently wait for something to arrive in the receive buffer, checking the URXDA flag.

3. Fetch the character from the receive buffer (FIFO).

Again, all of these steps can be nicely packaged in one last function:

```
char getU2( void)
{
    RTS = 0;                        // assert Request To Send !RTS
    while ( !U2STAbits.URXDA);      // wait for a new character to arrive
    return U2RXREG;                 // read the character from the receive buffer
    RTS = 1;
}// getU2
```

Testing the serial communication routines

To test our serial port control routines, we can now write a small program that will initialize the serial port, send a prompt, and let us type on the terminal keyboard while echoing each character back to the terminal screen:

```
main()
{
    char c;

    // 1. init the UART2 serial port
    initU2();

    // 2. prompt
    putU2( '>');

    // 3. main loop
    while ( 1)
    {

        // 3.1 wait for a character
        c = getU2();

        // 3.2 echo the character
        putU2( c);

    } // main loop

}// main
```

Follow these steps:

1. Build the project first, and then follow the standard checklist to activate the ICD2 Debugger and to program the Explorer16.

2. Connect the serial cable to the PC (directly or via a serial-to-USB converter) and configure HyperTerminal for the same communication parameters: 115200, n, 8, 1, RTS/CTS on the available COM port.

3. Click on HyperTerminal Connect button to start the terminal emulation.

4. Select "Run" from the Debugger menu to execute the demonstration program. Note: I recommend, for now, you do not attempt to single-step, use breakpoints, or RunToCursor when using the UART! See the "Tips and Tricks" section at the end of the chapter for a detailed explanation.

Note also that, if HyperTerminal is already set to provide an echo for each character sent, you will see double...literally! To disable this functionality, first hit the "Disconnect" button on HyperTerminal. Then select "File→Properties" and in the properties dialog box select the "Settings Pane Tab." This will be a good opportunity to set a couple more options that will come handy in the rest of the lesson.

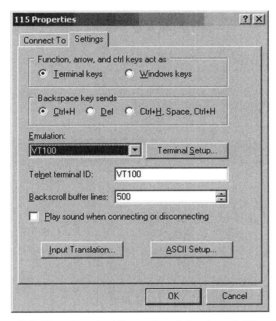

Figure 8-4. HyperTerminal Properties dialog box, Setting Pane.

1. Select the VT100 terminal emulation mode so that a number of commands (activated by special "escape" strings) will become available and will give us more control of the cursor position on the terminal screen.

2. Select ASCII Setup to complete the configuration. In particular, make sure that the "Echo typed characters locally" function is NOT checked (this will immediately improve your... vision). (See Figure 8-5.)

3. Also check the "Append line feeds to incoming line ends" option. This will make sure that every time an ASCII carriage return ('\r') character is received, an additional line feed ('\n') character is inserted automatically.

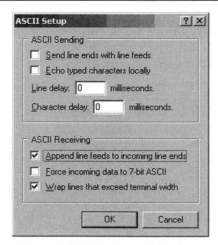

Figure 8-5. ASCII Setup dialog box.

Building a simple console library

To transform our demo project into a proper terminal console library that could become handy in future projects, we need only a couple more functions that will complete the puzzle: a function to print an entire (zero terminated) string and a function to input a full text line. Printing a string is, as you can imagine, the simple part:

```
int putsU2( char *s)
{
    while( *s)            // loop until *s == '\0', end of string
        putU2( *s++);     // send the character and point to the next one
} // putsU2
```

It is just a loop that keeps calling the `putU2` function to send, one after the other, each character in the string to the serial port.

Reading a text string from the terminal (console) into a string buffer can be equally simple, but we have to make sure that the size of the buffer is not exceeded (should the user type a really long string) and we have to convert the carriage-return character at the end of the line into a proper `'\0'` character for the string termination.

```
char *getsnU2( char *s, int len)
{
    char *p = s;          // copy the buffer pointer
    do{
        *s = getU2();     // wait for a new character

        if ( *s=='\r')    // end of line, end loop
            break;
        s++;              // increment buffer pointer
    len--;
    } while ( len>1 );    // until buffer full
```

```
    *s = '\0';                 // null terminate the string

    return p;                  // return buffer pointer
} // getsnU2
```

In practice the function, as presented, would prove very hard to use. There is no echo of what is being typed and the user has no room for errors. Make only the smallest typo and the entire line must be retyped. If you are like me, you do make a lot of typos…all of the time…, and the most battered key on your keyboard is the backspace key. A better version of the `getsnU2` function must include character echo and at least provisions for the backspace key to perform basic editing. It really takes only a couple more lines of code. The echo is quickly added after each character is received. The backspace character (identified by the ASCII code 0x8) is decoded to move the buffer pointer one character backward (as long as we are not at the beginning of the line already). We must also output a specific sequence of characters to visually remove the previous character from the terminal screen:

```
char *getsnU2( char *s, int len)
{
    char *p = s;               // copy the buffer pointer
    int cc = 0;                // character count
    do{
        *s = getU2();          // wait for a new character
        putU2( *s);            // echo character

        if (( *s==BACKSPACE)&&( s>p))
        {
            putU2( ' ');       // overwrite the last character
            putU2( BACKSPACE);
            len++;
            s--;               // back the pointer
            continue;
        }
        if ( *s=='\n')         // line feed, ignore it
            continue;
        if ( *s=='\r')         // end of line, end loop
            break;
        s++;                   // increment buffer pointer
        len--;
    } while ( len>1 );         // until buffer full

    *s = '\0';                 // null terminate the string

    return p;                  // return buffer pointer
} // getsnU2
```

Put all the functions in a separate file that we will call "conU2.c". Then create a small header file "conU2.h", to decide which functions (prototypes) and which constants to publish and make visible to the outside world:

```
/*
** CONU2.h
** console I/O library for Explorer16 board
*/

// I/O definitions for the Explorer16
#define CTS       _RF12        // Clear To Send, input, HW handshake
#define RTS       _RF13        // Request To Send, output, HW handshake
#define BACKSPACE 0x8          // ASCII backspace character code

// init the serial port (UART2, 115200@32MHz, 8, N, 1, CTS/RTS )
void initU2( void);

// send a character to the serial port
int putU2( int c);

// wait for a new character to arrive to the serial port
char getU2( void);

// send a null terminated string to the serial port
int putsU2( char *s);

// receive a null terminated string in a buffer of len char
char * getsnU2( char *s, int n);
```

Testing a VT100 terminal

Since we have enabled the VT100 terminal-emulation mode (see HyperTerminal settings above), we now have a few commands available to better control the terminal screen and cursor position, such as:

- clrscr, to clear the terminal screen.

- home, to move the cursor to the home position in the upper left corner of the screen.

These commands are performed by sending so called "escape sequences" (defined in the ECMA-48 standard, (also ISO/IEC 6429 and ANSI X3.64), also referred to as ANSI escape codes. They all start with the characters **ESC** (ASCII 0x1b) and the character '[' (left-squared bracket):

```
// useful macros for VT100 terminal emulation
#define clrscr() putsU2( "\x1b[2J")
#define home()   putsU2( "\x1b[1,1H")
```

In order to test the console library, we can now write up a small program that will:

1. Initialize the serial port.

2. Clear the terminal screen.

3. Send a welcome message/banner.

4. Send a prompt character.

5. Read a full line of text.

6. Print the text on a new line.

Save the following code in a new file that we will call "CONU2test.c":

```
/*
** CONU2 Test
** UART2 RS232 asynchronous communication demonstration code
*/

#include <p24fj128ga010.h>
#include "conU2.h"

#define BUF_SIZE 128

main()
{
    char s[BUF_SIZE];

    // 1. init the console serial port
    initU2();

    // 2. text prompt
    clrscr();
    home();
    putsU2( "Learn to fly with the PIC24!");

    // 3. main loop
    while ( 1)
    {
        putU2(">");      // prompt

      // 3.1 read a full line of text
        getsnU2( s, BUF_SIZE);

      // 3.2 send a string to the serial port
        putsU2( s);

      // 3.3 send a carriage return
      putU2('\r');

    } // main loop

}// main
```

Follow these steps:

1. Create a new project, using the "New Project" checklist, and add all three files: "conU2.h", "conU2.c" and "conU2test.c" to the project and build all.

2. Use the ICD2 checklist to connect the ICD2 debugger and program the Explorer16 board.

3. Test the editing capabilities of the new console library you just completed.

Using the serial port as a debugging tool

Once you have a small library of functions to send and receive data to a console through the serial port, you have a new and powerful debugging tool available. You can strategically position calls to print functions to present the content of critical variables and other diagnostic information on the terminal. You can easily format the output so as to be in the most convenient format for you to read. You can add input functions to set parameters that can better help test your code or you can use the input function to simply pause the execution and give you time to read the diagnostic output when required. This is one of the oldest debugging tools, effectively used since the first computer was ever invented.

The matrix

To finish this lesson on a more entertaining note, let's develop a new demo project that we will call the "matrix.c". The intent is to test the speed of the serial port and the PC terminal emulation by sending large quantities of text to the terminal and clocking its performance. The only problem is that we don't have (yet) access to a large storage device from which to extract some meaningful content to send to the terminal. So the next best option is that of "generating" some content using a pseudo-random number generator. The "stdlib.h" library offers in fact a convenient rand() function that returns a positive integer between 0 and MAX_RAND (a constant defined in the "limits.h" file that in the MPLAB C30 implementation can be verified to be equal to 32,767).

Using the "remainder of" operator we can reduce its output to any smaller integer range and produce only a subset of printable character values from the ASCII set. The following statement, for example, will produce only characters in the range from 33 to 127.

```
putU2( 33 + (rand()%94));
```

To generate a more appealing and entertaining output, especially if you happened to watch the movie *The Matrix*, we will present the (random) content by columns instead of rows. We will use the pseudo-random number generator to change the content and the "length" of each column as we continuously refresh the screen.

```
/*
**   The Matrix
**
*/
#include <p24fj128ga010.h>

#include "CONU2.h"
#include <stdlib.h>

#define COL     40
#define ROW     23
```

```
#define DELAY 3000

main()
{
    int v[40];   // vector containing length of each string
    int i,j,k;

    // 1. initializations
    T1CON = 0x8030;   // TMR1 on, prescale 256, Tcy/2
    initU2();         // initialize the console (115200, 8, N, 1, CTS/RTS)
    clrscr();         // clear the terminal (VT100 emulation)
    getU2();          // wait for one character to randomize the sequence
    srand( TMR1);

    // 2. init each column length
    for( j =0; j<COL; j++)
        v[j] = rand()%ROW;

    // 3. main loop
    while( 1)
    {
        home();

        // 3.1 refresh the screen with random columns
        for( i=0; i<ROW; i++)
        {
            // refresh one row at a time
            for( j=0; j<COL; j++)
            {
                // print a random character down to each column length
                if ( i < v[j])
                    putU2( 33 + (rand()%94));
                else
                    putU2(` `);
                putU2( ` `);
            } // for j
            pcr();
        } // for i

        // 3.2 randomly increase or reduce each column length
        for( j=0; j<COL; j++)
        {
            switch ( rand()%3)
            {
```

```
            case 0: // increase length
                    v[j]++;
                    if (v[j]>ROW)
                        v[j]=ROW;
                    break;

            case 1: // decrease length
                    v[j]--;
                    if (v[j]<1)
                        v[j]=1;
                    break;

            default:// unchanged
                    break;
        } // switch
    } // for


    } // main loop
} // main
```

Forget the performance—watching this code run is fun. It is too fast anyway—in fact, you will have to add a small delay loop (inside the for loop in 3.1) to make it more pleasant on the eye:

```
        // 3.1.1 delay to slow down the screen update
        TMR1 =0;
        while( TMR1<DELAY);
```

Note: remember to take the blue pill the next time!

Post-flight briefing

In this lesson we have developed a small console I/O library while reviewing the basic functionality of the UART module for operation as an RS232 serial port. We connected the Explorer16 board to a VT100 (emulated) terminal (Windows HyperTerminal). We will take advantage of this library in the next few lessons to provide us with a new debugging tool and possibly as a user interface for more advanced flights/projects.

Notes for C experts

I am sure at this point you are wondering about the possibility of using the more advanced library functions defined in the `"stdio.h"` library (such as `printf`) to direct the output to the UART2 peripheral. This is in fact possible by simply replacing one of the essential library functions: `"write.c"`:

```c
/*
** write.c
** replaces stdio lib write function
**
*/

#include <p24fj128ga010.h>
#include <stdio.h>
#include "conu2.h"

int write(int handle, void *buffer, unsigned int len)
{
    int i, *p;
    const char *pf;

    switch (handle)
    {
    case 0: // stdin
    case 1: // stdout
    case 2: // stderr
        for (i = len; i; --i)
            putU2( *(char*)buffer);
        break;
    default:
        break;
    } // switch
    return(len);
} // write
```

Save this code in a file called `"write.c"` in your project directory and add it to the list of source files for the project.

From this moment on, the linker will perform the connection and any call to one of the `"stdio.h"` library functions producing output on one of the standard streams (`stdin`, `stdout`, `stderr`) will be redirected to the UART2.

Notice that you will still be responsible for the proper UART initialization and the `"conu2.c"` file will have to be included in the project sources as well.

Notes for PIC microcontroller experts

Sooner or later, every embedded-control designer will have to come to terms with the USB bus. If, for now, a small "dongle" (converting the serial port to a USB port) can be a reasonable solution, sooner or later you are going to find opportunities and designs that will actually benefit from the superior performance and compatibility of the USB bus. Several 8-bit PIC microcontroller models already incorporate a USB Serial Interface Engine (SIE) as a standard communication interface. Microchip offers a free USB software stack with drivers and ready-to-use solutions for the most common classes of applications. One of them, known as the Communication Device Class (or CDC), makes the USB connection look completely transparent to the PC application so that even HyperTerminal cannot tell the difference. Most importantly, you will not need to write and/or install any special Windows drivers. When writing the application in C, you won't even notice the difference, if not for the absence of a need to specify any communication parameter. In USB there is no baud rate to set, no parity to calculate, no port number to select (wrong), while the communication speed is so much higher...

Tips and tricks

About the ICD2 and UARTs on ICE

As we mentioned during one of the early exercises presented in this lesson, single-stepping through a routine that enables and uses the UART to transmit and receive data from the HyperTerminal program is a bad idea. You will be frustrated, seeing the HyperTerminal program misbehave and/or simply lock up and ignore any data sent to it without any apparent reason. In order to understand the problems, you need to know more about how the MPLAB ICD2 in-circuit debugger operates. After executing each instruction, when in single-step mode or, upon encountering a breakpoint, the ICD2 debugger not only stops the CPU execution, but also "freezes" all the peripherals. It freezes them as in dead-cold-ice all of a sudden—not a single clock pulse is transmitted through their digital veins. When this happens to a UART peripheral that is busy in the middle of a transmission, the output serial line (TX) is also frozen in the current state. If a bit was being shifted out in that precise instant, and specifically if it was a 1, the TX line will be held in the "break" state (low) indeterminately.

The HyperTerminal program, on the other side, would sense this permanent "break" condition and interpret it as a line error. It will assume the connection is lost and it will disconnect. Since HyperTerminal is a pretty "basic" program, it will not bother letting you know what is going on...it will not send a beep, not an error message, nothing—it will just lock up!

If you are aware of the potential problem, this is not a big deal. When you restart your program with the ICD2, you will have just to remember to hit the HyperTerminal Disconnect button first and then the Connect button again. All operations will resume normally.

Exercises

1. Write a console library with buffered I/O (using interrupts) to minimize the impact on program execution (and debugging).

Books

*   Eady, F. (2005)

    Implementing 802.11 with Microcontrollers:
    Wireless Networking for Embedded Systems Designers

    Newnes, Burlington, MA

    Fred brings his humor and experience in embedded programming to make even wireless networking easy.

*   Axelson, J. (1999)

    USB Complete, 3rd ed.

    Lakeview Research, Madison, WI

    Jan's book has reached the third edition already. She has added more material at every step and still managed to keep things very simple.

Links

*   *http://en.wikipedia.org/wiki/ANSI_escape_code*

    This is a link to the complete table of ANSI escape codes as implemented by the VT100 HyperTerminal emulation.

Glass bliss

In the old days, big round instruments that looked like steam gauges populated the cockpit of every airplane, from the smallest single-engine Cessna to the ultrasonic Concord. Being so ubiquitous, the six principal instruments, placed always in the same order, had gained the affectionate nickname of the six-pack. But the next time you get on a commercial plane, peek into the cabin if you can. Sure, there are still plenty of knobs and switches, but right in front of the pilots you will notice there has been a big change. There is a large and flat piece of glass (or two). And "glass" is what the pilots call this revolution, although there is much more silicon behind it than most of them would suspect. It is the digital revolution of the cockpit, and it has happened only in the last few years.

Numerous powerful microprocessors work hard behind that glass to cram as much information as possible into a very simple, intuitive and possibly pleasing interface. Global positioning system (GPS) technology has been the driving force behind this innovation, and every airplane manufacturer today offers several advanced glass cockpit options for new models. Some are even speculating that the recent increase in sales of new airplanes, and the stimulus to the entire industry that has followed, might be attributed to the excitement generated by the new "glass cockpit."

Unfortunately, these are not exactly the type of airplanes that you, as a student pilot, would be flying for the first few lessons. It might take a little while for modern new airplanes to hit the schools fleet, but it is just a matter of time now—glass bliss is on the horizon.

The embedded world also makes copious use of glass, with LCD displays. Let's explore the basics of LCD interfaces…

Flight plan

In this lesson, we will learn how to interface with a small and inexpensive LCD display module. This project will be a good excuse for us to learn and use the Parallel Master Port (PMP), a new and flexible parallel interface available on the PIC24 microcontrollers.

Pre-flight checklist

In addition to the usual software tools, including the MPLAB® IDE, MPLAB C30 compiler and the MPLAB SIM simulator, this lesson will require only the use of the Explorer16 demonstration board and the MPLAB ICD2 In-Circuit Debugger.

The flight

The Explorer16 board can accommodate three different types of dot-matrix, alphanumeric LCD display modules and one type of graphic LCD display module. By default, it comes with a simple "2-rows by 16-character" display, a 3V alphanumeric LCD module (Tianma TM162JCAWG1) compatible with the industry-standard HD44780 controllers. These LCD modules are complete display systems composed of the LCD glass, column and row multiplexing drivers, power-supply circuitry and an intelligent controller, all assembled together into the so-called Chip On Glass (COG) technology. Thanks to this high level of integration, the circuitry required to control the dot-matrix display is greatly simplified. Instead of the hundreds of pins required by the column-and-row drivers to directly control each pixel, we can interface to the module with a simple 8-bit parallel bus using just eleven I/Os.

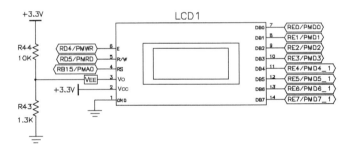

Figure 9-1. Default alphanumeric LCD Module connections.

On alphanumeric modules in particular, we can directly place ASCII character codes into the LCD module controller RAM buffer (DDRAM). The output image is produced by an integrated character generator (a table) using a 5 × 7 grid of pixels to represent each character. The table typically contains an extended ASCII character set, in the sense that it has been somewhat merged with a small subset of Japanese Kanji characters as well some symbols of common use. While the character generator table is mostly implemented in the display controller ROM, various display models offer the possibility to extend the character set by modifying/creating new characters (up to 8 on some models) accessing a second small internal RAM buffer (CGRAM).

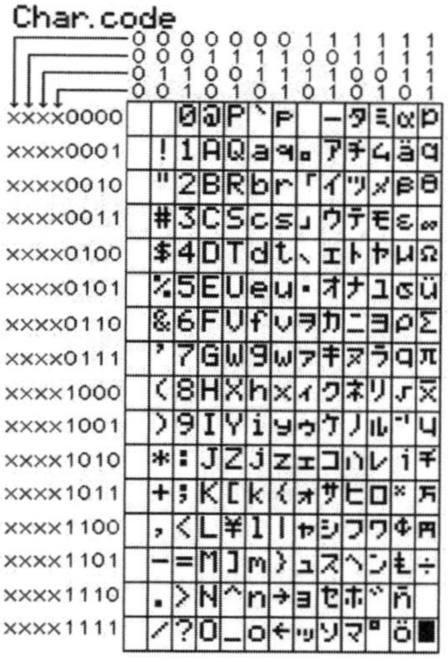

Figure 9-2. Character Generator table used by HD44780 compatible LCD display controllers.

HD44780 controller compatibility

As mentioned above, the 2 × 16 LCD module used in the Explorer16 board is one among a vast selection of LCD display modules available on the market in configurations ranging from one to four lines of 8, 16, 20, 32 and up to 40 characters each, that are compatible with the original HD44780 chipset, today considered an industry standard.

The HD44780 compatibility means that the integrated controller contains just two registers separately addressable, one for ASCII data and one for commands, and the following standard set of commands can be used to set up and control the display:

Instruction	Code										Description	Execution time
	RS	R/W	DB7	DB6	DB5	DB4	DB3	DB2	DB1	DB0		
Clear display	0	0	0	0	0	0	0	0	0	1	Clears display and returns cursor to the home position (address 0).	1.64ms
Cursor home	0	0	0	0	0	0	0	0	1	*	Returns cursor to home position (address 0). Also returns display being shifted to the original position. DDRAM contents remains unchanged.	1.64ms
Entry mode set	0	0	0	0	0	0	0	1	I/D	S	Sets cursor move direction (I/D), specifies to shift the display (S). These operations are performed during data read/write.	40us
Display On/Off control	0	0	0	0	0	0	1	D	C	B	Sets On/Off of all display (D), cursor On/Off (C) and blink of cursor position character (B).	40us
Cursor/display shift	0	0	0	0	0	1	S/C	R/L	*	*	Sets cursor-move or display-shift (S/C), shift direction (R/L). DDRAM contents remains unchanged.	40us

129

Instruction	Code										Description	Execution time
	RS	R/W	DB7	DB6	DB5	DB4	DB3	DB2	DB1	DB0		
Function set	0	0	0	0	1	DL	N	F	*	*	Sets interface data length (DL), number of display line (N) and character font(F).	40us
Set CGRAM address	0	0	0	1	CGRAM address						Sets the CGRAM address. CGRAM data is sent and received after this setting.	40us
Set DDRAM address	0	0	1	DDRAM address							Sets the DDRAM address. DDRAM data is sent and received after this setting.	40us
Read busy-flag and address counter	0	1	BF	CGRAM / DDRAM address							Reads Busy-flag (BF) indicating internal operation is being performed and reads CGRAM or DDRAM address counter contents (depending on previous instruction).	0us
Write to CGRAM or DDRAM	1	0	write data								Writes data to CGRAM or DDRAM.	40us
Read from CGRAM or DDRAM	1	1	read data								Reads data from CGRAM or DDRAM.	40us

Table 9-1. The HD44780 instruction set.

Bit name	Setting / Status	
I/D	0 = Decrement cursor position	1 = Increment cursor position
S	0 = No display shift	1 = Display shift
D	0 = Display off	1 = Display on
C	0 = Cursor off	1 = Cursor on
B	0 = Cursor blink off	1 = Cursor blink on
S/C	0 = Move cursor	1 = Shift display
R/L	0 = Shift left	1 = Shift right
DL	0 = 4-bit interface	1 = 8-bit interface
N	0 = 1/8 or 1/11 Duty (1 line)	1 = 1/16 Duty (2 lines)
F	0 = 5x7 dots	1 = 5x10 dots
BF	0 = Can accept instruction	1 = Internal operation in progress

Table 9-2. HD44780 command bits.

Thanks to this commonality, any code we will develop to drive the LCD on the Explorer16 board will be immediately available for use with any of the other HD44780-compatible alphanumeric LCD display modules.

The Parallel Master Port

The simplicity of the 8-bit bus shared by all these display modules is remarkable. Beside the eight bidirectional data lines (that could be reduced to just four for further I/O savings by enabling a special "nibble" mode), there is:

- An Enable strobe line (E).

- A Read/Write selection line (R/W̄) .

- An address line (RS) for the register selection.

It would be simple enough to control the 11 I/Os by manually controlling (bit-banging) the individual PORTE and PORTD pins to implement each bus sequence, but we will take this opportunity instead to explore the capabilities of a new peripheral introduced with the PIC24 architecture: the Parallel Master Port (PMP). The designers of the PIC24 family have created this new addressable parallel port to automate and accelerate access to a large number of external parallel devices of common use ranging from analog-to-digital converters, RAM buffers, ISA bus compatible interfaces, LCD display modules and even hard disks and CompactFlash® cards.

You can think of the PMP as a sort of flexible I/O bus added to the PIC24 architecture that does not interfere with (or slow down) the operation of the 24-bit wide program memory bus, nor the 16-bit data memory bus. The PMP offers:

- 8- or 16-bit bidirectional data path.

- Up to 64k of addressing space (16 address lines).

- Six additional strobe/control lines including:
  - Enable
  - Address Latch
  - Read
  - Write
  - and two Chip Select lines.

The PMP can also be configured to operate in slave mode, to attach as an addressable peripheral to a larger microprocessor/microcontroller system.

Both bus-read and bus-write sequences are fully programmable so that not only the polarity and choice of control signals can be configured to match the target bus, but also the timing can be finely tuned to adapt to the speed of the peripherals we interface to.

Configuring the PMP for LCD module control

As in all other PIC24 peripherals, there is a set of control registers dedicated to the PMP configuration. The first one is PMCON, and you will recognize the familiar sequence of control bits common to all the module xxCON registers.

Upper Byte:							
R/W-0	U-0	R/W-0	R/W-0	R/W-0	R/W-0	R/W-0	R/W-0
PMPEN	—	PSIDL	ADRMUX1	ADRMUX0	PTBEEN	PTWREN	PTRDEN
bit 15							bit 8

Lower Byte:							
R/W-0	R/W-0	R/W-0[1]	R/W-0[1]	R/W-0[1]	R/W-0	R/W-0	R/W-0
CSF1	CSF0	ALP	CS2P	CS1P	BEP	WRSP	RDSP
bit 7							bit 0

Figure 9-3. PMCON control register.

But the list of control registers that we will need to initialize is a bit longer this time and includes also: PMMODE, PMADDR, PMSTAT, PMPEN and possibly PADCFG1. They are packed with powerful options and they all require your careful consideration. Instead of proceeding through the lengthy review of each and every one of them, I will list only the key choices required specifically by the LCD module interface:

- PMP enabled
- Fully demultiplexed interface (separate data and address lines will be used)
- Enable strobe signal enabled (RD4)
- Read signal enabled (RD5)
- Enable strobe active high
- Read active high, Write active low
- Master mode with Read and Write signals on the same pin (RD5)
- 8-bit bus interface (using PORTE pins)
- Only one address bit is required, so we will choose the minimum configuration including PMA0 (RB15) and PMA1

Also, considering that the typical LCD module is an extremely slow device, we better select the most generous timing, adding the maximum number of wait states allowed at each phase of a read or write sequence:

- 4 × Tcy wait data set-up before read/write
- 15 × Tcy wait between R/W and Enable

* 4 × Tcy wait data set-up after Enable.

A small library of functions to access an LCD display

Create a new project using the "New Project" checklist and a new source file.

We will start writing the LCD initialization routine first. It is natural to start with the initialization of the PMP port key control registers:

```
void LCDinit( void)
{
    // PMP initialization
    PMCON = 0x83BF;         // Enable the PMP, long waits
    PMMODE = 0x3FF;         // Master Mode 1
    PMPEN = 0x0001;         // PMA0 enabled
```

After these steps we are able to communicate with the LCD module for the first time and we can follow a standard LCD initialization sequence recommended by the manufacturer. The initialization sequence must be timed precisely; see the HD44780 instruction set for the details. It cannot be initiated before at least 30 ms have been granted to the LCD module to proceed with its own internal initialization (power-on reset) sequence. For simplicity and safety, we will hard code a delay in the LCD module initialization function and we will use Timer1 to obtain simple but precise timing loops for all subsequent steps:

```
    // init TMR1
    T1CON = 0x8030;             // Fosc/2, prescaled 1:256, 16us/tick

    // wait for >30ms
    TMR1 = 0; while( TMR1<2000);    // 2000 x 16us = 32ms
```

For our convenience, we will also define a couple of constants that will help us, hopefully, make the following code more readable:

```
#define LCDDATA 1               // RS = 1 ; access data register
#define LCDCMD  0               // RS = 0 ; access command register
#define PMDATA  PMDIN1          // PMP data buffer
```

To send each command to the LCD module, we will select the command register (setting the address PMA0 = RS = 0) first. Then we will start a PMP write sequence by depositing the desired command byte in the PMP data output buffer:

```
    PMADDR = LCDCMD;            // select the command register (ADDR = 0)

    PMDATA = 0b00111000;       // function set: 8-bit interface, 2 lines, 5x7
```

The PMP will perform the complete bus write sequence as listed below:

1. The address will be published on the PMP address bus (PMA0).

2. The content of PMDATA will be published on the PMP data bus (PMD0-PMD7).

3. After 4 × Tcy the R/$\overline{W}$ signal will be asserted low (RD5).

4. After 15 × Tcy the Enable strobe will be asserted high (RD4).

5. After 4 × Tcy the Enable strobe will lowered and PMDATA removed from the bus.

Notice how this sequence is quite long as it extends for more than 20 × Tcy or more than 1.25 µs after the PIC24 has initiated it. In other words the PMP will still be busy executing part of this sequence while the PIC24 will have already executed another 20 instructions or more. Since we are going to wait for a considerably longer amount of time anyway (>40 µs) to allow the LCD module to execute the command, we will not worry about the time required by the PMP to complete the command at this time:

```
TMR1 = 0; while( TMR1<3);    // 3 x 16us = 48us
```

We will then proceed similarly with the remaining steps of the LCD module initialization sequence:

```
PMDATA = 0b00001100;         // display ON, cursor off, blink off
TMR1 = 0; while( TMR1<3);    // 3 x 16us = 48us

PMDATA = 0b00000001;         // clear display
TMR1 = 0; while( TMR1<100);  // 100 x 16us = 1.6ms

PMDATA = 0b00000110;         // increment cursor, no shift
TMR1 = 0; while( TMR1<100);  // 100 x 16us = 1.6ms
```

After the LCD module initialization, things will get a little easier and the timing loops will not be necessary anymore, as we will be able to use the LCD module Read Busy Flag command. This will tell us if the integrated LCD module controller has completed the last command given and is ready to receive and process a new one. In order to read the LCD status register containing the Busy Flag, we will need to instruct the PMP to execute a bus read sequence. This is a two-step process: we initiate the read sequence by reading (and discarding) the contents of the PMP data buffer a first time. When the PMP sequence is completed, the data buffer will contain the actual value read from the bus, and we will read its contents from the PMP data buffer again. But how can we tell when the PMP read sequence is complete? Simple—we can check the PMP Busy flag in the PMSTAT control register.

In summary, to check the LCD module Busy flag we will need to check the PMP Busy flag first, issue a read command, wait for the PMP Busy flag again, and finally we will gain access to the actual LCD module status-register contents, including the LCD Busy flag.

By passing the register address as a parameter to the read function, we will obtain a more generic function that will be able to read the LCD status register or the data register as in the following code:

```
char LCDread( int addr)
{
    int dummy;
    while( PMMODEbits.BUSY);    // wait for PMP to complete previous commands
    PMADDR = addr;              // select the command address
    dummy = PMDATA;             // initiate a read cycle, dummy read
    while( PMMODEbits.BUSY);    // wait for PMP to complete the sequence
    return( PMDATA);            // read the status register
} // LCDread
```

The LCD module status register contains two pieces of information: the LCD Busy flag and the LCD RAM pointer current value. We can use two simple macros to split the two pieces: LCDbusy() and LCDaddr(), and a third one to access the data register: getLCD():

```
#define LCDbusy()  LCDread( LCDCMD) & 0x80
#define LCDaddr()  LCDread( LCDCMD) & 0x7F
#define getLCD()   LCDread( LCDDATA)
```

Using the `LCDbusy()` function, we can create a function to write data or commands to the LCD module:

```
void LCDwrite( int addr, char c)
{
    while( LCDbusy());
    while( PMMODEbits.BUSY);     // wait for PMP to be available
    PMADDR = addr;
    PMDATA = c;
} // LCDwrite
```

A few additional macros will help complete the library:

*   `putLCD()` will send ASCII data to the LCD module:

    ```
    #define putLCD( d)  LCDwrite( LCDDATA, (d))
    ```

*   `LCDcmd()` will send generic commands to the LCD module:

    ```
    #define LCDcmd( c)  LCDwrite( LCDCMD, (c))
    ```

*   `LCDhome()` will reposition the cursor on the first character of the first row:

    ```
    #define LCDhome()   LCDwrite( LCDCMD, 2)
    ```

*   `LCDclr()` will clear the entire contents of the display:

    ```
    #define LCDclr()    LCDwrite( LCDCMD, 1)
    ```

And finally, for our convenience, we might want to add `putsLCD()`, a function that will send an entire null terminated string to the display module:

```
void putsLCD( char *s)
{
    while( *s)
        putLCD( *s++);
} //putsLCD
```

Let's put all of this to work adding a short main function:

```
main( void)
{
    // initializations
    LCDinit();

    // put a title on the first line
    putsLCD( "Flying the PIC24");

    // main loop, empty for now
     while ( 1)
     {
     }
} // main
```

If all went well after building the project and programming the Explorer16 board with the ICD2 debugger (using the usual checklists), you will now have the great satisfaction of seeing the title string published on the first row of the LCD display.

Advanced LCD control

If you felt that all of the preceding was not too complex, and certainly not rewarding enough, here I have some more interesting stuff, and a new challenge for you to consider.

When introducing the HD44780-compatible alphanumeric LCD modules, we mentioned how the display content was generated by the module controller by using a table, the character generator, located in ROM. But we also mentioned the possibility of extending the character set using an additional RAM buffer (known as the CGRAM). Writing to the CGRAM it is possible to create new 5 × 7 character patterns to create new symbols and possibly small graphic elements.

How about adding a small airplane to the character set of the Explorer16 LCD module display?

We will need a function to set the LCD module RAM buffer pointer to the beginning of the CGRAM area using the "Set CGRAM Address" command or, better, a macro that uses the LCDwrite() function:

```
#define LCDsetG( a) LCDwrite( LCDCMD, (a & 0x3F) | 0x40)
```

To generate two new 5 × 7 character patterns, one for the nose of the plane and one for the tail, we will use the putLCD() function. Each byte of data will contribute 5 bits (lsb) to define one row of the pattern. After the last row of each character is defined, an extra byte of data (8th) will be inserted to align for the next character block.

```
// generate two new characters
LCDsetG(0);
putLCD( 0b00010);
putLCD( 0b00010);
putLCD( 0b00110);
putLCD( 0b11111);
putLCD( 0b00110);
putLCD( 0b00010);
putLCD( 0b00010);
putLCD( 0);      // alignment

putLCD( 0b00000);
putLCD( 0b00100);
putLCD( 0b01100);
putLCD( 0b11100);
putLCD( 0b00000);
putLCD( 0b00000);
putLCD( 0b00000);
putLCD( 0);      // alignment
```

The two new symbols will now be accessible with the codes 0 and 1 respectively of the character-generator table.

To reposition the buffer pointer back to the data RAM buffer (DDRAM), use the following macro:

```
#define LCDsetC( a) LCDwrite( LCDCMD, (a & 0x7F) | 0x80)
```

Notice that while the first line of the display corresponds to addresses from 0 to 0xf of the DDRAM buffer, the second line is always found at addresses from 0x40 to 0x4f independently of the display size —the number of characters that compose each line of the actual display.

Also a simple delay mechanism (based once more on Timer1) will be necessary to make sure that our airplane flies on time and stays visible. LCD displays are slow, and if the display is updated too fast the image tends to disappear like a ghost:

```
#define TFLY 9000        // 9000 x 16us = 144ms
#define DELAY() TMR1=0; while( TMR1<TFLY)
```

It is time to devise a simple algorithm to make the little airplane fly in the main loop. Here it is:

```
// main loop
while( 1)
{
    // the entire plane appears at the right margin
    LCDsetC(0x40+14);
    putLCD( 0); putLCD( 1);
    DELAY();

    // fly fly fly (right to left)
    for( i=13; i>=0; i--)
    {
        LCDsetC(0x40+i);        // set the cursor to the next position
        putLCD(0); putLCD(1);   // new airplane
        putLCD(' ');            // erase the previous tail
        DELAY();
    }

    // the tip disappears off the left margin, only the tail is visible
    LCDsetC(0x40);
    putLCD( 1); putLCD(' ');
    DELAY();

    // erase the tail
    LCDsetC(0x40);                   // point to the left margin of the 2nd line
    putLCD(' ');

    // and draw just the tip appearing from the right
    LCDsetC(0x40+15);                // point to the right margin of the 2nd line
    putLCD( 0);
    DELAY();

} // repeat the main loop
```

Have fun flying the PIC24!

Post-flight briefing

In this lesson we learned how to use the Parallel Master Port to drive an LCD display module. Actually, we have just started scratching the surface. Also, since the LCD display module is a relatively slow peripheral, it might seem that there has been little or no significant advantage in using the PMP instead of a traditional bit-banged I/O solution. In reality, even when accessing such simple and slow peripherals the use of the PMP can provide two important benefits:

- the timing, sequence and multiplexing of the control signals is always guaranteed to match the configuration parameters, eliminating the risk of dangerous bus collisions and/or unreliable operation as a consequence of coding errors and/or unexpected execution and timing conditions (interrupts, bugs, ...).

- the MCU is completely free from tending to the peripheral bus, allowing simultaneous execution of any number of higher priority tasks.

Notes for C experts

As we did in the previous lesson, when using the asynchronous serial interfaces, it is possible to replace the low-level I/O routines defined in the "stdio.h" library, and in particular "write.c", to redirect the output to the LCD display. We can actually extend the previous example by providing redirection to the UART2 for the standard streams (stdin, stdout and stderr) and adding a fourth stream for the LCD as in the following example code:

```c
/*
** write.c
** replaces stdio lib write function
**
*/

#include <p24fj128ga010.h>
#include <stdio.h>

#include "conU2.h"

#include "LCD.h"

int write(int handle, void *buffer, unsigned int len)
{
    int i, *p;
    const char *pf;

    switch (handle)
    {
    case 0: // stdin
    case 1: // stdout
    case 2: // stderr
            for (i = len; i; --i)
                putU2( *(char*)buffer);
```

138

```
                break;

        case LCD: // additional stream
                for (i = len; i; --i)
                        putLCD( *(char*)buffer);
                break;
        default:
                break;
        } // switch
        return(len);
} //write
```

In an alternate scheme, you might want to redirect the "stdout" stream to the LCD display as the main output of the application, and the "stderr" stream to the serial port for debugging purposes.

Also, it is likely that at this point, you will want to modify the putLCD() function to interpret special characters like '\n', to advance to the next line, or even to introduce a partial decoding for a few ANSI escape codes so in order to be able to position the cursor and clear the screen (using the macros defined in this lesson) just as on a terminal console.

Tips and tricks

Since the LCD display is a slow peripheral, waiting for its commands to be completed in tight (locking) loops as in the examples provided in this lesson could constitute an unacceptable waste of MCU cycles in some applications. A better scheme would require caching LCD commands in a FIFO buffer and using an interrupt mechanism to periodically schedule their execution. In other words, interrupts would be used to perform multitasking of a slow process in the background of the application execution.

An example of such a mechanism is provided in the "LCD.c" example code provided with the Explorer16 demonstration board.

Exercises

1. Enhance the putLCD() function to correctly interpret the following characters:

    –   `'\n'`: advance to the next line.

    –   `'\r'`: reposition cursor to the beginning of current line.

    –   `'\t'`: advance to a fixed tabulation position.

2. Enhance the putLCD() function to interpret the following ANSI escape codes:

    –   `'\x1b[2J'`: clear entire screen.

    –   `'\x1b[1,1H'`: home cursor.

    –   `'\x1b[n,mH'`: position the cursor at row `'n'`, column `'m'`.

Books

*   Bentham, J.

    TCP/IP Lean, Web Servers for Embedded Systems

    CMP Books, Lawrence, Kansas

    This book will take you one level of complexity higher, showing you how the TCP/IP protocols, the foundation of the Internet, can be easily implemented in a "few" lines of C code. Jeremy knows how to keep things "lean," as is necessary in every embedded-control application.

Links

*   *http://www.microchip.com/stellent/idcplg?IdcService=SS_GET_PAGE&nodeId= 1824&appnote=en011993*

    This is a link to Microchip Application Note 833, a free TCP/IP stack for all PICmicros.

*   *http://www.microchip.com/stellent/idcplg?IdcService=SS_GET_PAGE&nodeId= 1824&appnote=en012108*

    Application Note 870 describes a Simple Network Management Protocol for Microchip TCP/IP stack-based applications.

It's an analog world

There are certain things that, no matter how many times you practice, never seem to come out the same way twice. Landings are a good example. Even the most experienced airline captains will occasionally have a bad day and screw it up. I'm sure you must have noticed it, when they "bounce" a landing. What is wrong with landings? Why are they so difficult?

The fact is that, no matter how hard you try, the conditions affecting a landing are never really *exactly* the same. The wind speed and direction change continuously, the performance of the engine changes, and even the wings change with the slightest change in the air temperature. Additionally, the pilot reflexes (and alertness) change. It all combines to create an infinite number of unpredictable conditions that make for an infinite number of possible ways to get it wrong.

We live in an analog world. All the input variables, temperature, wind speed and direction are analog. All of our sensory system inputs (sight, sounds, pressure) are analog. The output, such as the movement of the pilot's muscles to control the plane, is analog. With time, we learn to interpret (or should I say convert) all the analog inputs from the world around us and make the best decisions we can. Practice can make us perfect, almost!

In embedded control, the information from the analog world must first be converted to digital. The analog-to-digital converter module is one of the key interfaces to the "real" world.

Flight plan

The PIC24 family was designed with embedded-control applications in mind and therefore is ideally prepared to deal with the analog nature of the world. A fast analog-to-digital converter (ADC), capable of 500,000 conversions per second, is available on all models with an input multiplexer that allows you to monitor a number of analog inputs quickly and with high resolution. In this lesson we will learn how to use the 10-bit ADC module available on the PIC24FJ128GA010 family to perform two simple measurements on the Explorer16 board: reading a voltage input from a potentiometer first and a voltage input from a temperature sensor later.

Preflight checklist

In addition to the usual software tools, including the MPLAB® IDE, MPLAB C30 compiler and the MPLAB SIM simulator, this lesson will require the use of the Explorer16 demonstration board and the MPLAB ICD2 In-Circuit Debugger.

The flight

The first step in using the analog-to-digital converter, as with any other peripheral modules inside the PIC24, is to familiarize yourself with the module building blocks and the key control registers. Yes, this means reading the datasheet once more, and even the Explorer16 User Guide to look at the schematics. We can start by looking at the ADC module block diagram:

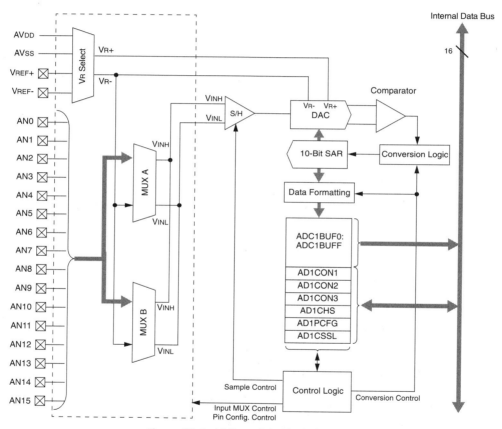

Figure 10-1. ADC module block diagram

This is a pretty sophisticated structure that offers many interesting capabilities:

- Up to 16 input pins can be used to receive the analog inputs.

- Two input multiplexers can be used to select different input analog channels and different reference sources each.

- The output of the 10-bit converter can be formatted for integer or fixed-point arithmetic, and signed or unsigned 16-bit output.

- The control logic allows for many possible automated conversion sequences so as to synchronize the process to the activity of other related modules and inputs

- The conversion output is stored in a 16-bit wide, 16-words deep buffer that can be configured for sequential scanning or simple FIFO buffering.

All these capabilities require a number of control registers to be properly configured and I understand how, especially at the beginning, the number of options available and decisions to take could make you a bit dizzy. So we will start by taking the shortest and simplest approach with the simplest example application: reading the position of the R6 potentiometer on the Explorer16 board.

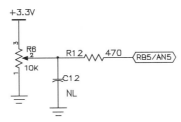

Figure 10-2. Detail of the Explorer16 demonstration board, R6 potentiometer.

The 10 kohm potentiometer is connected directly to the power-supply rails so that its output can span the entire range of values from 3.3V to the ground reference. It is connected to the RB5 pin that corresponds to the analog input AN5 of the ADC input multiplexer.

After creating a new project using the appropriate checklist, we can create a new source file "pot.c" including the usual header file and adding the definition of a couple of useful constants. The first one, POT, defines the input channel assigned to the potentiometer and the second one, AINPUTS, is a mask that will help us define which inputs should be treated as analog and which ones as digital:

```
/*
** It's an analog world
** Converting the analog signal from a potentiometer
*/

#include <p24fj128ga010.h>

#define POT     5       // 10k potentiometer connected to AN5 input
#define AINPUTS 0xffef  // Analog inputs for Explorer16 POT and TSENS
```

The actual initialization of all the ADC control registers can be best performed by a short function, `initADC()`, that will produce the desired initial configuration:

- `AD1PCFG` will be passed the mask selecting the analog input channels: 0s will mark the analog inputs, and 1s will configure the respective pins as digital inputs.

- `AD1CON1` will set the conversion to start automatically, triggered by the completion of the auto-timed sampling phase; also, the output will be formatted for a simple unsigned, right-aligned (integer) value.

- `AD1CSSL` will be cleared, as no scanning function will be used (only one input).

- `AD1CON2` will select the use of MUXA and will connect the ADC reference inputs to the analog input rails AVdd and AVss pins.

- `AD1CON3` will select the conversion clock source and divider.

- Finally, setting `ADON`, the entire ADC peripheral will be activated and ready for use.

```
void initADC( int amask)
{
    AD1PCFG = amask;              // select analog input pins
    AD1CON1 = 0;                  // manual conversion sequence control
    AD1CSSL = 0;                  // no scanning required
    AD1CON2 = 0;                  // use MUXA, AVss and AVdd are used as Vref+/-
    AD1CON3 = 0x1F02;             // Tad = 2 x Tcy = 125ns >75ns
    AD1CON1bits.ADON = 1;         // turn on the ADC
} //initADC
```

By passing `amask` as a parameter to the initialization routine, we make it flexible enough to accept multiple input channels in future applications.

The first conversion

The actual analog-to-digital conversion is a two-step process. First we need to take a sample of the input voltage signal, and then we can disconnect the input and perform the actual conversion of the sampled voltage to a numerical value. The two distinct phases are controlled by two separate control bits in the `AD1CON1` register: `SAMP` and `DONE`. The timing of the two phases is important to provide the necessary accuracy of the measurement:

- During the sampling phase the external signal is connected to an internal capacitor that needs to be charged up to the input voltage. Enough time must be provided for the capacitor to track the input voltage and this time is mainly proportional to the impedance of the input signal source (in our case known to be less than 5 kohm) as well as the internal capacitor value. In general, the longer the sampling time, the better the result, compatibly with the input signal frequency (not an issue in our case).

- The conversion phase timing depends on the selected ADC clock source. This is typically derived by the main CPU clock signal via a divider or alternatively by an independent RC oscillator. The RC option, although appealing for its simplicity, is a good choice when a conversion needs to be performed during a sleep (low-power mode) phase, when the CPU clock is turned off. The oscillator clock divider is a better option in more general cases, since it provides synchronous operation with the CPU and therefore a better rejection of the noise internally produced by it. The conversion clock should be the fastest possible, compatibly with the specifications of the ADC module (in our case Tad is required to be longer than 75 ns, requiring a minimum clock divider by two).

Here is a basic conversion routine:

```
int readADC( int ch)
{
    AD1CHS  = ch;                       // 1. select analog input channel

    AD1CON1bits.SAMP = 1;               // 2. start sampling

    TMR1 = 0;                           // 3. wait for sampling time
    while (TMR1< 100);                  // 6.25 us

    AD1CON1bits.DONE = 1;               // 4. start the conversion

    while (!AD1CON1bits.DONE);          // 5. wait for the conversion to complete

    return ADC1BUF0;                    // 6. read the conversion result

} // readADC
```

Automatic sampling timing

As you can see, using this basic method, we have been responsible for providing the exact timing of the sampling phase, dedicating a timer to this task and performing two waiting loops. But on the PIC24 there is a new option that allows for a more automatic process. The sampling phase can be self timed, provided the input source impedance is small enough to require a maximum sampling time of $32 \times$ Tad (32×120 ns = 3.8 µs in our case). This can be achieved by setting the SSRC bits in the AD1CON1 register to the 0b111 configuration, so as to enable an automatic start of the conversion upon termination of a self-timed sampling period. The period itself is selected by the AD1CON3 register SAM bits. Here is a new and improved example that uses the self-timed sampling and conversion trigger:

```
void initADC( int amask)
{
    AD1PCFG = amask;            // select analog input pins
    AD1CON1 = 0x00E0;           // automatic conversion start after sampling
    AD1CSSL = 0;                // no scanning required
    AD1CON2 = 0;                // use MUXA, AVss and AVdd are used as Vref+/-
    AD1CON3 = 0x1F02;           // Tsamp = 32 x Tad; Tad=125ns
    AD1CON1bits.ADON = 1;       // turn on the ADC
} //initADC
```

Notice how making the conversion-start be triggered automatically by the completion of the self-timed sampling phase gives us two advantages:

- Proper timing of the sampling phase is guaranteed without requiring us to use any timed delay loop and/or other resource.

- One command (start of the sample phase) suffices to complete the entire sampling and conversion sequence.

With the ADC so configured, starting a conversion and reading the output is a trivial matter:

- AD1CHS selects the input channel for MUXA.

- Setting the SAMP bit in AD1CON1 starts the timed sampling phase, which will be immediately followed by the conversion.

- The DONE bit will be set in the AD1CON1 register as soon as the entire sequence is completed and a result is ready.

- Reading the ADC1BUF0 register will immediately return the desired conversion result.

```c
int readADC( int ch)
{
    AD1CHS   = ch;              // 1. select analog input channel

    AD1CON1bits.SAMP = 1;      // 2. start sampling

    while (!AD1CON1bits.DONE);  // 3. wait for the conversion to complete

    return ADC1BUF0;           // 4. read the conversion result

} // readADC
```

Developing a demo

All that remains to do at this point is to figure out an entertaining way to put the converted value to use on the Explorer16 demo board. The LEDs connected to PORTA are an obvious choice, but instead of simply providing a binary output, publishing the eight most significant bits of the 10-bit result, why not jazz things up a little and provide a visual feedback more reminiscent of the analog nature of our input? We could turn on one LED at a time, using it as an index on a mechanical dial. Here is the main routine we will use to test our analog-to-digital functions:

```c
main ()
{
    int a;

    // initializations
    initADC( AINPUTS);  // initialize the ADC for the Explorer16 analog inputs
    TRISA = 0xff00;     // select the PORTA pins as outputs to drive the LEDs

    // main loop
    while( 1)
    {
```

```
        a = readADC( POT);  // select the POT input and convert

        // reduce the 10-bit result to a 3 bit value (0..7)
        // (divide by 128 or shift right 7 times
        a >>= 7;

        // turn on only the corresponding LED
        // 0 -> leftmost LED.... 7-> rightmost LED
        PORTA = (0x80 >> a);

    } // main loop
} // main
```

After the call to the initialization routine (to which we provide a mask that defines bit 5 as analog input), we initialize the TRISA register to make the pins connected to the LED bar digital outputs. Then, in the main loop we perform the conversion on AN5 and we reformat the output to fit our special display requirements. As configured, the 10-bit conversion output will be returned as a right-aligned integer in a range of values between 0 and 1024. By dividing that value by 128 (or in other words shifting it right seven times) we can reduce the range to a 0 to 7 value. The final output, though, requires one more transformation to produce the eight desired LED configurations. Note that the LED corresponding to the MSB is located to the left of the bar, and to maintain the correspondence between the potentiometer movement clockwise and the index LED moving to the right we need to start with a 0b10000000 pattern and shift it right as required.

Build the project and, following the usual ICD2 debugging checklist, program the Explorer16 board. If all goes well, you will be able to play with the potentiometer, moving it from side to side while observing the LED index moving left and right correspondingly.

Developing a game

OK, I will admit it, the previous example was not too exciting. After all, we have been using a 16-MIPS capable 16-bit machine to perform an analog-to-digital conversion roughly 200,000 times a second (32 Tad sampling + 12 Tad conversion, where Tad = 125 ns, you do the math) only to discard all but three bits of the result and watch a single LED light up. How about making it a bit more challenging and playful instead? How about developing a little "Whac-A-Mole"[1] game, just a mono-dimensional version?

Let's turn on a second LED (the mole), controlled by the PIC24 and distinguishable from the user-controlled LED (the mallet) because it's somewhat dimmer. By moving the mallet (bright LED), rotating the potentiometer until you reach the mole (dim LED), you will get to "whack it"! A new mole, in a different random position will immediately appear and the game will continue.

The pseudo-random number generator function rand() (defined in "stdlib.h") will be helpful here, as all (computer) games need a certain degree of unpredictability. We will use it to determine where to place each new mole.

---

[1] Whac-a-Mole is a trademark of Bob's Space Racers Inc.

Save the source file from the first project with a new name "LEDgame.c" and create an entire new project. Then update the main() function to include just a few more lines of code:

```
main ()
{
    int a, r, c;

    // 1. initializations
    initADC( AINPUTS);  // initialize the ADC for the Explorer16 analog inputs
    TRISA = 0xff00;     // select the PORTA pins as outputs to drive the LEDs

    // 2. use the first reading to randomize the number generator
    srand( readADC( POT));
    r = 0x80;
    c = 0;

    // 3. main loop
    while( 1)
    {
        a = readADC( POT);  // select the POT input and convert

        // 3.1  reduce the 10-bit result to a 3 bit value (0..7)
        // (divide by 128 or shift right 7 times
        a >>= 7;

        // 3.2 turn on only the corresponding LED
        // 0 -> leftmost LED.... 7-> rigtmost LED
        a = (0x80 >> a);

        // 3.3 as soon as the cursor hits the random dot, generate a new one
        while (a == r )
            r = 0x80 >> (rand() & 0x7);

        // 3.4 display the user (bright) LED and food (dim) LED
        if ((c & 0xf) == 0)
            PORTA = a + r;  // add food LED only 1/16 of the times (dim)
        else
            PORTA = a;      // always display the user LED (bright)

        // 3.5 loop counter
        c++;

    } // main loop
} // main
```

- In 1, we perform the usual initialization of the analog-to-digital converter module and the PORTA I/Os connected to the bar of LEDs.

- In 2, we read the potentiometer value for the first time and we use its position as the SEED value for the random-number generator. This makes the game experience truly unique each time, provided the potentiometer is not always found in the leftmost or rightmost position. That would provide a SEED value of 0 or 1023, respectively, every time and therefore would make the game quite repetitive as the pseudo-random sequence would proceed through the same steps at any game restart.

- In 3, the main loop begins, as in the previous example, reading an integer 10-bit value and reducing it to the three most significant bits. (3.1).

- In 3.2, the conversion into an LED position "a" is performed just as before, but it is in 3.3 that things get interesting. If the position of the user LED represented by "a" is overlapping the "mole" LED position "r", a new random position is immediately calculated. The operation needs to be repeated as a while loop because, each time a new random value for "r" is calculated, there is a chance (exactly 1/8 if our pseudo-random generator is a good one) that the new value could be the same. In other words we could be creating a new "mole" right under the mallet. And that would not be very challenging or sporting. Don't you agree?

- Steps 3.4 and 3.5 are all about displaying and differentiating the two LEDs. To show both LEDs on the display bar, we could simply "add" the two binary patterns "a" and "r" but it would be very hard for the player to tell which is which. To represent the "mole" LED with a dimmer light, we can alternate cycles of the main loop where we present both LEDs and cycles where only the "mallet" LED is visible. Since the main loop is executed hundreds of thousands of time per second, our eye will perceive the "mole" LED as dimmer, proportionally to the number of cycles it is missing. For example, if we add the "mole" LED only once every 16 cycles, its apparent brightness will be only 1/16 that of the "mallet" LED.

- The counter "c", constantly incremented in 3.5, helps us to implement this mechanism.

- In 3.4 we look only at the 4 lsb of the counter (0...15) and we add the "mole" LED to the display only when their value is 0b0000. For the remaining 15 loops, only the "mallet" LED will be added to the display.

Build the project and download it to the Explorer16 board. You have to admit that it's much more entertaining now!

Measuring temperature

Moving on to more serious things, there is a temperature-sensor mounted on the Explorer16 board and it happens to be a Microchip TC1047A integrated temperature-sensing device with a nice linear voltage output. This device is very small, as it is offered in a SOT-23 (three-pin, surface-mount) package. The power consumption is limited to 35 µA (typ.) while the power supply can cover the entire range from 2.5V to 5.5V. The output voltage is independent from the power supply and is an extremely linear function of the temperature (typically within 0.5 degree C) with a slope of exactly 10 mV/C. The offset is adjusted to provide an absolute temperature indication according to the formula seen in Figure 10-3.

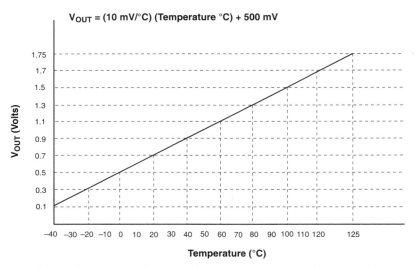

$$V_{OUT} = (10 \text{ mV/}°C) (\text{Temperature } °C) + 500 \text{ mV}$$

Figure 10-3. TC1047 Output Voltage vs. Temperature characteristics.

We can apply our newly acquired abilities to convert the voltage output to digital information using, once more, the analog-to-digital converter of the PIC24. The temperature sensor is directly connected to the AN4 analog input channel as per the Explorer16 board schematic.

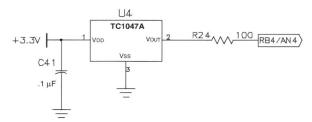

Figure 10-4. Detail of the Explorer16 demonstration board, TC1047A temperature sensor.

We can reuse the ADC functions developed for the previous exercise and put them in a new project called "TSense" and save the previous source file as `"Tsense.c"`.

Let's start modifying the code to include a new constant definition: TSENS for the ADC input channel assigned to the temperature sensor:

```
/*
** It's an analog world
** Converting the analog signal from a TC1047 Temperature Sensor
*/

#include <p24fj128ga010.h>
```

```
#define POT      5        // 10k potentiometer connected to AN5 input
#define TSENS    4        // TC1047 Temperature sensor with voltage output
#define AINPUTS 0xffcf    // Analog inputs for Explorer16 POT and TSENS

// initialize the ADC for single conversion, select Analog input pins
void initADC( int amask)
{
    AD1PCFG = amask;     // select analog input pins
    AD1CON1 = 0x00E0;    // auto convert after end of sampling
    AD1CSSL = 0;         // no scanning required
    AD1CON3 = 0x1F02;    // max sample time = 31Tad, Tad = 2 x Tcy = 125ns >75ns
    AD1CON2 = 0;         // use MUXA, AVss and AVdd are used as Vref+/-
    AD1CON1bits.ADON = 1; // turn on the ADC
} //initADC

int readADC( int ch)
{
    AD1CHS  = ch;               // select analog input channel
    AD1CON1bits.SAMP = 1;       // start sampling, auto-conversion will follow
    while (!AD1CON1bits.DONE);  // wait to complete the conversion
    return ADC1BUF0;            // read the conversion result
} // readADC
```

As you can see, nothing else needed to change with regard to the ADC configuration or activation of the conversion sequence. Presenting the result on the LED bar might be a little tricky though. Temperature sensors provide a certain level of noise and, to give a more stable reading, it is common to perform a little filtering. Taking groups of 16 samples and performing an average will give us a cleaner value to work with:

```
a = 0;
for ( j= 16; j >0; j--)
    a += readADC( TSENS);  // add up 16 successive temperature readings
i = a >> 4;                // divide the result by 16 to obtain the average
```

But how could we display the result using only the bar of LEDs?

We could pick the most significant bits of the conversion result and publish them in binary or BCD, but once more it would not be fun. How about providing instead a relative temperature indication using a similar (single LED) index moving along the LED bar?

We will sample the initial temperature value just before the main loop and use it as the offset for the center bar position. In the main loop we will update the dot position, moving it to the right as the sensed temperature increases or to the left as the sensed temperature decreases. Here is the complete code for the new temperature-sensing example:

```
main ()
{
    int a, i, j;
```

```
// 1. initializations
initADC( AINPUTS);  // initialize the ADC for the Explorer16 analog inputs
TRISA = 0xff00;     // select the PORTA pins as outputs to drive the LEDs
T1CON =   0x8030;   // TMR1 on, prescale 1:256 Tclk/2

// 2. sample initial temp value
a = 0;
for ( j= 16; j >0; j--)
    a += readADC( TSENS);  // read the temperature
i = a >> 4;
// this will give the central bar reference

// 3. main loop
while( 1)
{
    // 3.1 read a new (averaged) temperature value
    a = 0;
    for ( j= 16; j >0; j--)
    {
        TMR1 = 0;
        while ( TMR1 < 3900);  // 3900 x 256 x Tcy ~= 1sec
        a += readADC( TSENS);  // read the temperature
    }
    a >>= 4;                    // averaged over 16 readings

    // 3.2 compare with the initial reading and move the bar 1 pos. per C
    a = 3 + (a - i);

    // 3.3 keep the result in the value range 0..7, keep the bar visible
    if ( a > 7)
        a = 7;
    if ( a < 0)
        a = 0;

    // 3.4 turn on the corresponding LED
    PORTA = ( 0x80 >> a);

} // main loop
} // main
```

- In 3.2, we determine the difference between the initial reading "i" and the new averaged reading "a". The result is centered, so that a central LED is lit up when the difference is zero.

- In 3.3 the result is checked against the boundaries of the available display. Once the difference becomes negative and more than three bits wide, the display must simply indicate the leftmost position. When the difference is positive and more than four bits wide, the rightmost LED must be activated.

- In 3.4 we publish this result as in the previous example.

To complete the exercise and give you a more aesthetically pleasing experience, I recommend that you also introduce an additional delay loop (for convenience inserted inside the 3.1 averaging loop). This will slow things down quite a bit, reducing the update rate of the display (and eventually the entire main loop cycle) to a period of about one second. A faster update rate would produce only an annoying flicker when the temperature readings are too close to the intermediate values between two contiguous dot positions.

Build the project with the usual checklists and download it to the Explorer16 board.

After identifying the temperature sensor on the board (hint: it is close to the lower left corner of the PIC24 processor module and it looks like a surface-mount transistor), run the program and observe how small temperature variations, obtained by touching or blowing hot/cold air on the sensor, move the cursor around.

The breath-alizer game

To have a bit more fun with the temperature sensor, we can now merge the last two exercises into one new game. We'll call it the "Breath-alizer" game. The idea is to whack the "mole" (dim) LED by controlling the "mallet" using the temperature sensor. Heat the sensor up with some hot air to move it to the right, blow cold air on it to move it to the left. Have fun!

```
main ()
{
    int a, i, j, k, r;

    // 1. initializations
    initADC( AINPUTS);  // initialize the ADC for the Explorer16 analog inputs
    TRISA = 0xff00;     // select the PORTA pins as outputs to drive the LEDs
    T1CON = 0x8030;     // TMR1 on, prescale 1:256 Tclk/2

    // 2. use the first reading to randomize the number generator
    srand( readADC( TSENS));
    // generate the first random position
    r = 0x80 >> (rand() & 0x7);
    k = 0;

    // 3. compute the average value for the initial reference
    a = 0;
    for ( j= 16; j >0; j--)
        a += readADC( TSENS);  // read the temperature
    i = a >> 4;

    // 5. main loop
    while( 1)
    {
        // 5.1 take the average value over 1 second
        a = 0;
        for ( j= 16; j >0; j--)
```

```
        {
            TMR1 = 0;
            while ( TMR1 < 3900)   // 16 x 3900 x 256 x Tcy ~= 1sec
            { // display the user LED and dim random LED
                if ((TMR1 & 0xf) == 0)
                    PORTA = k + r;
                else
                    PORTA = k ;
            }

            a += readADC( TSENS);   // read the temperature
        }
        a >>= 4;                         // averaged over 16 readings

        // 5.2 compare with the initial reading and move the bar 1 pos. per C
        a = 3 + (a - i);
        // keep the result in the value range 0..7, keep the bar visible
        if ( a > 7)
            a = 7;
        if ( a < 0)
            a = 0;
        // update the user LED
        k = ( 0x80 >> a);

        // 5.3 as soon as the user hits the random LED, generate a new position
        while (k == r )
            r = 0x80 >> (rand() & 0x7);

    } // main loop
} // main
```

Post-flight briefing

In this lesson with have just started scratching the surface and exploring the possibilities offered by the analog-to-digital converter module of the PIC24. We have used one simple configuration of the many possible and only a few of the advanced features available. We have tested our newly acquired capabilities with two types of analog inputs available on the Explorer16 board, and hopefully we had some fun in the process.

Notes for C experts

Even if the PIC24 has a fast divide instruction, there is no reason to waste any processor cycles. In embedded control, "every" processor cycle is precious. If the divisor is a power of two, the integer division can be best performed as a simple shift right by an appropriate number of positions with a computational cost that is at least an order of magnitude smaller than a regular division. If the divider is not a power of two, consider changing it if the application allows. In our last example, we could have opted for averaging 10 temperature samples, or 15 as well as 20, but we chose 16 because this made the division a simple matter of shifting the sum by 4 bits to the right (in a single cycle PIC24 instruction).

Tips and tricks

If the sampling time required is longer than the maximum available option (32 × Tad) you can try to extend Tad first, or a better option is to swap things around and enable the automatic sampling start (at the end of the conversion). This way the sampling circuit is always open, charging, whenever the conversion is not occurring. Manually clearing the SAMP bit will trigger the actual conversion start.

Further, having Timer3 periodically clearing the SAMP control bit for you (one of the options for the SSRC bits in AD1CON1), and enabling the ADC end of conversion interrupt will provide the widest choice of sampling periods possible for the least amount of MCU overhead possible. No waiting loops, only a periodic interrupt when the results are available and ready to be fetched.

Exercises

1. Use the ADC FIFO buffer to collect conversion results; set up Timer 3 for automatic conversion and the interrupt mechanism so that a call is performed only once the buffer is full and temperature values are ready to be averaged.

Books

* Baker, B.

   A Baker's Dozen: Real Analog Solutions for Digital Designers

   Newnes, Burlington, MA

   For proper care and feeding of an analog-to-digital converter, look no further than Bonnie's cookbook.

Links

* *http://www.microchip.com/stellent/idcplg?IdcService=SS_GET_PAGE&nodeId=2102&param =en021419&pageId=79&pageId=79*

Temperature sensors are available in many flavors and with a choice of interface options, including direct I²C™ or SPI digital output.

Cross-Country Flying

Part III – Cross-country flying

Congratulations! You have endured a few more lessons and gained the ability to complete more complex flights. You are now going to enter the third and last part of your training where you'll practice cross-country flying. No more pattern work around the airport, no more landings and take-offs, or maneuvers in the practice area—you will finally get to go somewhere!

In the third part of this book, we will start developing new projects that will require you to master several peripheral modules at once. Since the examples will become a bit more complex, not only having an actual demonstration board (the Explorer16) at hand is recommended, but also having the ability to perform small modifications and utilize the prototyping area to add new functionality to the demonstration board will be necessary. Simple schematics and component part numbers will be offered in the following chapters as required. On the companion web site "FlyingthePIC24.com" (and/or "ProgrammingthePIC24.com") you will find additional expansion boards and prototyping options to help you enjoy even the most advanced projects.

Capturing inputs

As we were saying in a previous chapter, advanced electronics is rapidly gaining space in the cockpits of all but the smallest airplanes. While the glass (LCD) displays are supplanting the old steam gauges, GPS satellite receivers are plotting the airplane position in real time on colorful maps depicting terrain elevations and, with additional equipment, up-to-the-minute satellite weather information too. Pilots can enter an entire flight plan in the navigation system and then follow their path on the moving map, just like in a video game. The interaction with these new instruments, though, is becoming the next big challenge. Just as with computer applications, each instrument is controlled by a different menu system and a set of knobs and buttons to allow the pilot to provide the inputs quickly and, hopefully, intuitively. However, the limited space in the cockpit has so far imposed serious limitations on the type and number of such input devices, which for the most part—at least in the first generations—have been mimicking the knobs and buttons of the primitive VHF radios.

If you have a GPS navigation system in your car and you have tried to dial in a street address in a foreign city (say "Bahnhofstrasse, 17, Munich") by twisting and turning that little knob while driving on a highway…well, you know exactly the type of challenge I am talking about. Keyboards are the logical next level of input interface for several advanced avionics (aviation electronics) systems. They are already common in business jet cockpits, but they are starting to make their appearance in the smaller general aviation airplanes too. How about a keyboard in your next car?

Flight plan

With the advent of the USB bus, computers have finally been freed from a number of "legacy" interfaces that had been in use for decades since the introduction of the first IBM PC. The PS/2 mouse and keyboard interface is one of them. The result of this transition is that a large number of the "old" keyboards are now flooding the surplus market and even new PS/2 keyboards are selling for very low prices. This creates the opportunity to give our future PIC24 projects a powerful input capability. It will also give us the motivation to investigate a few alternative interface methods and their trade-offs. We'll implement software state machines, refresh our experience using interrupts and possibly learn to use new peripherals.

The flight

The physical PS/2 port uses a 5-pin DIN or a 6-pin mini-DIN connector. The first was common on the original IBM PC-XT and AT series but has not been in use for a while now. The smaller 6-pin version has been more common in recent years. Once the different pin-outs are taken into consideration, the two are electrically identical.

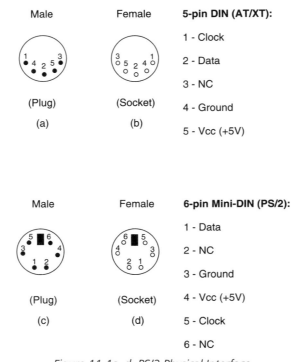

Figure 11-1a–d. PS/2 Physical Interface.

The host must provide a 5V power supply. The current consumption will vary with the keyboard model and year, but you can expect values between 50 and 100 mA (the original specifications used to call for up to 275 mA max).

The data and clock lines are both open-collector with pull-up resistors (1–10 kohm) so as to allow for two-way communication. In the normal mode of operation, it is the keyboard that drives both lines in order to send data to the personal computer. When it is necessary, though, the computer can take control to configure the keyboard and to change the status LEDs ("Caps" and "Num Lock").

The PS/2 communication protocol

At idle, both the data and clock lines are held high by the pull-ups (located inside the keyboard). In this condition the keyboard is enabled and can start sending data as soon as a key has been pressed. If the host holds the clock line low for more than 100 µs, any further keyboard transmissions are suspended. If the host holds the data line low and then releases the clock line, this is interpreted as a request to send a command.

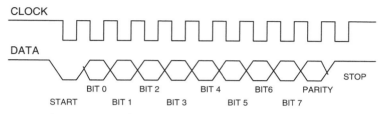

Figure 11-2. Keyboard to Host communication waveform.

The protocol is a curious mix of synchronous and asynchronous communication protocols that we have seen in previous chapters. It is synchronous since a clock line is provided, but it is similar to an asynchronous protocol since a start, a stop and a parity bit are used to bracket the actual 8-bit data packet. Unfortunately, the baud rate used is not a standard value and can change from unit to unit, over time, with temperature and the phase of the moon. In fact, typical values range from 10 to 16 kbit per second. Data changes during the clock-high state. Data is valid when the clock line is low. Whether data is flowing from the host to the keyboard or vice versa, it is the keyboard that always generates the clock signal.

Note: The USB bus reverses the roles as it makes each peripheral a synchronous slave of the host. This simplifies things enormously for a non-real-time, nonpreemptive multitasking operating system like Windows®. The serial port and the parallel port were similarly asynchronous interfaces, and—probably for the same reason—both became legacy with the introduction of the USB bus specification.

Interfacing a PIC24 to the PS/2

The unique peculiarities of the protocol make interfacing to a PS/2 keyboard an interesting challenge, as neither the PIC24 SPI interface nor the UART interface can be used. In fact, the SPI interface does not accept 11-bit words (8-bit or 16-bit words are the only options), while the PIC24 UART would require the periodic transmission of special break characters to make use of the powerful auto baud-rate detection capabilities. Also notice that the PS/2 protocol is based on 5V-level signals. This requires care in choosing which pins can be directly connected to the PIC24. In fact, only the 5V-tolerant digital input pins can be used, which excludes the I/O pins that are multiplexed with the analog-to-digital converter.

Input Capture

The first idea that comes to mind is to implement in software a PS/2 serial interface peripheral using the Input Capture mechanism.

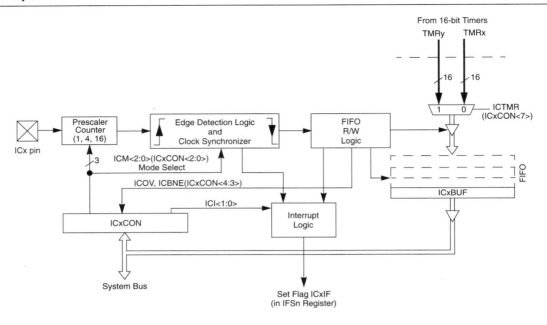

Note: An 'x' in a signal, register or bit name denotes the number of the capture channel.

Figure 11-3. Input Capture module block diagram.

Five Input Capture modules are available on the PIC24FJ128GA010, connected respectively to the IC1-IC5 pins multiplexed on PORTD pins 8, 9, 10, 11 and 12.

Each Input Capture module is controlled by a single corresponding control register ICxCON and works in combination with one of two timers, either Timer2 or Timer3.

One of several possible events can trigger the input capture:

- rising edge
- falling edge
- rising and falling edge
- 4th rising edge
- 16th rising edge.

The current value of the selected timer is recorded and stored in a FIFO buffer to be retrieved by reading the corresponding ICxBUF register. In addition to the capture event, an interrupt can be generated after a programmable number of events (each time, every second, every third or every fourth).

To put the Input Capture peripheral to use and receive the data stream from a PS/2 keyboard, we can connect the IC1 input to the clock line and configure the peripheral to generate an interrupt on each and every falling edge of the clock.

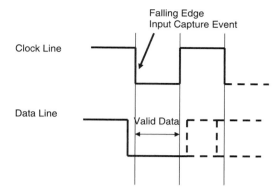

Figure 11-4. PS/2 Interface Bit Timing and the Input Capture trigger event.

After creating a new project, and following our usual template, we can start adding the following initialization code:

```
#define PS2DATA  _RG12          // any available 5V tolerant input
#define PS2CLOCK _RD8           // use the IC1 module input pin

void initKBD( void)
{
    // clear the flag
    KBDReady = 0;

    _TRISD8 = 1;                // make IC1 = RD8 pin an input (clock)
    _TRISG12 = 1;               // make the RG12 pin an input (data)
    IC1CON = 0x0002;            // use TMR3, int every capture, falling edge
    _IC1IF = 0;                 // clear the interrupt flag
    _IC1IE = 1;                 // enable the IC1 interrupt

} // void initKBD
```

We will also need to create an interrupt service routine for the IC1 interrupt vector. This routine will have to operate as a state machine and perform the following steps in sequence:

1. Verify the presence of a start bit (data line low).

2. Shift in 8 bits of data and compute a parity.

3. Verify a valid parity bit.

4. Verify the presence of a stop bit (data line high).

If any of these checks fails, the state machine must reset and return to the start condition. When a valid byte of data is received, we will store it in a buffer—think of it as a mailbox—and a flag will be raised so that the main program or any other "consumer" routine will know a valid key code has been received and is ready to be retrieved. To fetch the code, it will suffice to copy it from the mailbox first and then to clear the flag.

165

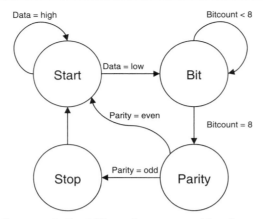

Figure 11-5. The PS/2 receive state machine diagram.

The state machine requires only four states and a counter, and all the transitions are summarized in Table 11-1:

State	Conditions	Effect
Start	Data = low	Init Bitcount, Init Parity, Transition to Bit state
Bit	Bitcount < 8	Shift in key code, LSB first (shift right), Update Parity Increment Bitcount
Bit	Bitcount = 8	Transition to Parity state
Parity	Parity = even	Error. Transition back to Start
Parity	Parity = odd	Transition to Stop
Stop	Data = low	Error. Transition back to Start
Stop	Data = high	Save the key code in buffer, Set flag, Transition to Start

Table 11-1. PS/2 receive state machine transitions table.

Theoretically we should consider this an 11-state machine, counting each time the Bit state is entered with a different Bitcount value as a distinct state. But the four-state model works best for an efficient C-language implementation. Let's define a few constants and variables that we will use to maintain the state machine:

```
// definition of the keyboard PS/2 state machine
#define PS2START    0
#define PS2BIT      1
#define PS2PARITY   2
#define PS2STOP     3

// PS2 KBD state machine and buffer
int PS2State;
unsigned char KBDBuf;              // temporary buffer
int KCount, KParity,;              // bitcount and parity

// key code flag and mailbox
volatile int KBDReady;
volatile unsigned char KBDCode;
```

Note: The keyword `volatile` is used as a modifier in a variable declaration to alert the compiler that the content of the variable could change unpredictably as a consequence of an interrupt or other hardware mechanism. We use it here to prevent the compiler from applying any optimization technique (loop extraction, procedure abstraction..) whenever these two variables are used. Admittedly, we could have omitted the detail in this code example (after all, all optimizations are supposed to be turned off during debugging), only to find ourselves with a big headache in the future, when using this code in a more complex project and trying to squeeze it to get the highest possible performance. `KBDReady` and `KBDcode` are the only two variables used in both the interrupt service routine and the main interface code.

The interrupt service routine for the input capture IC1 module can finally be implemented using a simple switch statement that performs the entire state machine.

```
void _ISR _IC1Interrupt( void)
{ // input capture interrupt service routine

    switch( PS2State){
        default:
        case PS2START:
            if ( ! PS2DAT)
            {
                KCount = 8;         // init bit counter
                KParity = 0;        // init parity check
                PS2State = PS2BIT;
            }
            break;
```

```
        case PS2BIT:
            KBDBuf >>=1;                // shift in data bit
            if ( PS2DAT)
                KBDBuf += 0x80;
            KParity ^= KBDBuf;          // update parity
            if ( --KCount == 0)         // if all bit read, move on
                PS2State = PS2PARITY;
            break;

        case PS2PARITY:
            if ( PS2DAT)
                KParity ^= 0x80;
            if ( KParity & 0x80)        // if parity is odd, continue
                PS2State = PS2STOP;
            else
                PS2State = PS2START;
            break;

        case PS2STOP:
            if ( PS2DAT)                // verify stop bit
            {
                KBDCode = KBDBuf;       // save the key code in mail box
                KBDReady = 1;           // set flag, key code available
            }
            PS2State = PS2START;
            break;

    } // switch state machine

    // clear interrupt flag
    _IC1IF = 0;

} // IC1 Interrupt
```

Testing the Input Capture method using Stimulus Scripts

The small perforated prototyping area can be used to attach a PS/2 mini-DIN connector to the Explorer16 demonstration board, the only alternative being the development of a custom daughter board (PICTail™) for the expansion connectors. Before committing to designing such a board though, we would like to make sure that the chosen pin-out and code is going to work. The MPLAB® SIM software simulator is going to be, once more, our tool of choice.

While in previous chapters we have used the software simulator in conjunction with the Watch window, the Stopwatch, and the Logic Analyzer to verify that our programs were generating the proper timings and outputs, this time we will need to simulate inputs as well. To this end MPLAB SIM offers a considerable number of options and resources, so many in fact that the system might seem a bit intimidating at first. First of all, the simulator offers two types of input stimuli: asynchronous ones, typically triggered manually by the user, and synchronous ones, triggered automatically by the simulator after a scripted amount of time (expressed in processor cycles or seconds). The script files (.SCL) containing the descriptions of the synchronous stimuli (that can be quite complex) can be prepared using a convenient tool, called the SCL Generator. You can invoke the SCL Generator by selecting "SCLGenerator→ New Workbook" from the Debugger menu. In order to prepare the simplest type of stimulus script, one that assigns values to specific input pins (and entire registers) at given points in time, you can select the first tab in the Generator window: "Pin/Register Actions".

After selecting the unit of measurement of choice, microseconds in our case, click on the first row of the table that occupies most of the dialog box window space (where it says "Click here to Add Signals"). This will allow you to add columns to the table. Add one column for every pin for which you want to simulate inputs. In our example, that would be RG12 for the PS2 Data line and IC1 for the Input Capture pin that we want connected to the PS2 Clock line. At this point we can start typing in the stimulus timing table. To simulate a generic PS/2 keyboard transmission, we will need to produce a 10-kHz clock signal for 11 cycles as represented in the PS/2 keyboard waveform in Figure 11-4. This requires an event to be inserted in the timing table every 50 μs. As an example, Table 11-2 illustrates the trigger events I recommend you add to the SCL Generator timing table to simulate the transmission of key code 0x79.

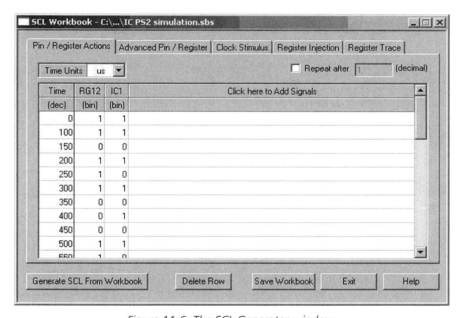

Figure 11-6. The SCL Generator window.

Time (us)	RG12	IC1	Comment
0	1	1	Idle state, both lines are pulled up
100	1	1	
150	0	0	First falling edge, Start bit (0)
200	1	1	
250	1	0	Bit 0, key code LSb (1)
300	0	1	
350	0	0	Bit 1 (0)
400	0	1	
450	0	0	Bit 2 (0)
500	1	1	
550	1	0	Bit 3 (1)
600	1	1	
650	1	0	Bit 4 (1)
700	1	1	
750	1	0	Bit 5 (1)
800	1	1	
850	1	0	Bit 6 (1)
900	0	1	
950	0	0	Bit 7, key code MSb (0)
1000	0	1	
1050	0	0	Parity bit (0)
1100	1	1	
1150	1	0	Stop bit (1)
1200	1	1	Idle

Table 11-2. SCL Generator timing example for basic PS/2 simulation.

Once the timing table is filled, you can save the current content for future use with the "Save Workbook" button. The file generated will be an ASCII file with the .SBS extension. In theory, you could edit this file manually with the MPLAB IDE editor or any basic ASCII editor, but you are strongly discouraged from doing so. The format is more rigid than meets the eye and you might end up trashing it. If you were wondering why the term "Workbook" is used for what looks like a simple table, you are invited to explore the other panes (accessible clicking on the tabs at the top of the dialog box) of the SCL Generator. You will see that what we are using in this example is just one of the many stimuli generation methods available, representing a minuscule portion of the capabilities of the SCL Generator. A Workbook file can contain a number of different types of stimuli produced by any (or several) of those panes.

A segment of the SCL Generator Workbook file is shown here:

```
## SCL Builder Setup File: Do not edit!!

## VERSION: 3.22.00.00
## FORMAT:  v1.40.00
## DEVICE:  PIC24FJ128GA010

## PINREGACTIONS
us
No Repeat
RG12
IC1
--
0
1
1
--
100
1
1
--
150
0
0
--
200
1
1
```

At this point an actual stimulus script file can be generated from the timing table we just defined. Stimulus script files have a `.scl` extension and are, once more, simple ASCII text files. The script files contain the real commands and information that the MPLAB SIM simulator will use to simulate the actual input signals. A segment of the stimulus file is shown here:

```
//
// .../IC PS2 simulation.scl
// Generated by SCL Generator ver. 3.22.00.00
// DATE TIME
//


configuration for "pic24fj128ga010" is
end configuration;


testbench for "pic24fj128ga010" is
begin
```

```
process is
begin
    wait for 0 us;
    report "Stimulus actions after 0 us";
    RG12 <= '1';
    IC1 <= '1';
    wait;
end process;

process is
begin
    wait for 100 us;
    report "Stimulus actions after 100 us";
    RG12 <= '1';
    IC1 <= '1';
    wait;
end process;
```

You might notice a certain resemblance between the notation used in the SCL file and some hardware description languages (VHDL). Perhaps it is not just a coincidence!

The structured format adopted is, in fact, designed to allow great flexibility in describing the stimuli as well as a fast simulation execution.

Testing the PS/2 receive routines

Before we get to use the stimulus file generated, we have to complete the project with a few final touches. Let's package the PS/2 receive routines as a module that we might want to call "PS2IC.c". Remember to include the file in the project (right-click in the editor window, and "Add to Project").

Let's also prepare an include file to publish the accessible function: initKBD(), the flag KBDReady and the buffer for the received key code KBDCode:

```
/*
**
**   PS2IC.h
**
** PS/2 keyboard input library using input capture
*/

extern volatile int KBDReady;
extern volatile unsigned char KBDCode;

void initKBD( void);
```

Note that there is no reason to publish any other detail of the inner workings of the PS2 receiver implementation. This will give us freedom to later try a few different methods without changing the interface. Save this file as "PS2IC.h" and include it in the project.

Let's also create a new file "PS2ICTest.c" that will contain the main routine and will use the PS2IC module to test its functionality:

```
/*
** PS2 KBD Test
**
*/

#include <p24fj128ga010.h>

#include "PS2IC.h"

main()
{
    TRISA = 0xff00;
    initKBD();                  // call the initialization routine

    while ( 1)
    {
        if ( KBDReady)          // wait for the flag
        {
            PORTA = KBDCode;    // fetch the key code and publish on PORTA
            KBDReady = 0;       // clear the flag
        }
    } // main loop
} //main
```

This will initialize PORTA LSB for output (on the Explorer16 connected to the LEDs), and will call the PS/2 keyboard initialization routine that, in its turn, will initialize the chosen input pins, the state machine, and the interrupts on input capture.

The main loop will wait for the interrupt routine to raise the flag (key code available), will fetch the key code and publish it on the LEDs, and finally will clear the flag, ready to receive a new character.

Now remember to add the file to the project and "Build All."

The simulation

Instead of launching the simulation immediately, proceed to the Debugger menu once more and select the "Stimulus Controller" submenu.

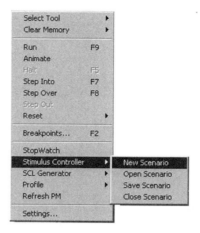

Figure 11-7. Stimulus Controller submenu.

Select "New Scenario" and you will see a new dialog box appear on the screen. This is the Stimulus Controller and, although it looks deceptively similar to the SCL Generator dialog box, don't let it fool you!

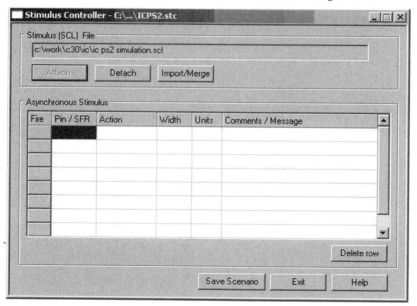

Figure 11-8. The Stimulus Controller window.

The Stimulus Controller allows you to attach to the project the synchronous stimulus scripts you generated with the SCL Generator, and add to them "asynchronous stimuli" triggered by the "Fire" buttons that you see in the Stimulus Controller table.

Select the "Attach" button and select the .SCL file we generated before.

You could now save this "scenario" for later use, but in our case, since we will be dealing with just this one .SCL file and no further asynchronous stimulus will be created, there is really no point.

> Note: You must keep the Stimulus Controller window open (in the background). Resist the temptation to hit the Exit button, as that would close the scenario and leave us without stimuli.

Finally! Hit the Reset button (or select "Debugger→Reset") and watch for the first stimulus to arrive as the microsecond 0 trigger is fired. Remember, both lines RG12 and IC1 are supposed to be set high according to our simulation timetable. A message will be confirming this in the Output window.

Figure 11-9. The output window (MPLAB SIM pane) showing that a stimulus action has been triggered.

It is your choice now to proceed by single-stepping or animating through the program to verify its correct execution. My suggestion is that you start by placing a breakpoint inside the main loop, just past the instruction copying the KBDCode on to PORTA. Open the Watch window and add PORTA from the SFR list, and then RUN.

After a few seconds, the execution should terminate at the breakpoint and the content of PORTA should reflect the data we sent through the simulated PS/2 lines: 0x79!

The Simulator Profile

If you were curious about how fast the simulation of a PIC24 could run on your computer, there is an interesting feature available to you in the MPLAB SIM Debugger menu, the Profile. Select the Profile submenu ("Debugger→Profile") and click on "Reset Profile" first. (See Figure 11-10.)

This will clear the Simulator Profile counters and timers. Then remove all breakpoints and let the simulator run ("Debugger→Run") for a few seconds. Halt the simulation and go back to the "Debugger→Profile" submenu. This time, select "Display Profile". (See Figure 11-11.)

A relatively long report will be available in the output window (MPLAB SIM pane), listing how many times each instruction was used by the processor during the simulation and at the very bottom offering an assessment of the absolute simulation speed. In my case, that turned out to be a respectable 2.7 MIPS, meaning the software simulation (on my laptop) ran at about one-sixth the actual processor speed. Not bad at all!

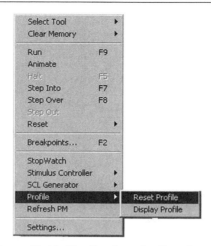

Figure 11-10. The Simulator Profile submenu.

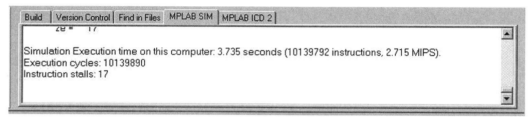

Figure 11-11. Simulator Profile output.

Another method – Change Notification

While the Input Capture technique worked all right, there are other options that we might be curious to explore in order to interface efficiently with a PS/2 keyboard. In particular, there is another interesting peripheral available in the PIC24 that could offer an alternative method to implement a PS/2 interface: the Change Notification (CN) module. There are as many as 22 I/O pins connected to this module and this can give us some freedom in choosing the ideal input pins for the PS/2 interface, while making sure they don't conflict with other functions required in our project or already in use on the Explorer16 board. There are only four control registers associated with the CN module. The CNEN1 and CNEN2 registers contain the interrupt-enable control bits for each of the CN input pins. Setting any of these bits enables a CN interrupt for the corresponding pins. Note that only one interrupt vector is available for the entire CN module, therefore it will be the responsibility of the interrupt service routine to determine which one of the enabled inputs has actually changed.

File Name	Addr	Bit 15	Bit 14	Bit 13	Bit 12	Bit 11	Bit 10	Bit 9	Bit 8	Bit 7	Bit 6	Bit 5	Bit 4	Bit 3	Bit 2	Bit 1	Bit 0	All Resets
CNEN1	0060	CN15IE	CN14IE	CN13IE	CN12IE	CN11IE	CN10IE	CN9IE	CN8IE	CN7IE	CN6IE	CN5IE	CN4IE	CN3IE	CN2IE	CN1IE	CN0IE	0000
CNEN2	0062	—	—	—	—	—	—	—	—	—	—	CN21IE	CN20IE	CN19IE	CN18IE	CN17IE	CN16IE	0000
CNPU1	0068	CN15PUE	CN14PUE	CN13PUE	CN12PUE	CN11PUE	CN10PUE	CN9PUE	CN8PUE	CN7PUE	CN6PUE	CN5PUE	CN4PUE	CN3PUE	CN2PUE	CN1PUE	CN0PUE	0000
CNPU2	006A	—	—	—	—	—	—	—	—	—	—	CN21PUE	CN20PUE	CN19PUE	CN18PUE	CN17PUE	CN16PUE	0000

Legend: — = unimplemented, read as '0'. Reset values are shown in hexadecimal.

Table 11-3. The CN control register table.

Each CN pin also has a weak pull-up connected to it. The pull-up acts as a current source that is connected to the pin and eliminates the need for external resistors when push-button or keypad devices are connected. The pull-ups are enabled separately using the CNPU1 and CNPU2 registers, which contain the control bits for each of the CN pins. Setting any of the control bits enables the weak pull-ups for the corresponding pins.

In practice, all we need to support the PS/2 interface is only one of the CN inputs connected to the PS2 clock line. The PIC24 weak pull-up will not be necessary in this case, as it is already provided by the keyboard.

There are 22 pins to choose from, and we will find a CN input that is not shared with the analog-to-digital converter (remember we need a 5V-tolerant input) and is not overlapping with some other peripheral used on the Explorer16 board. This takes a little studying between the device datasheet and the Explorer16 user guide. But once the input pin is chosen, say CN11 (multiplexed with PORTG pin 9, the SS line of the SPI2 module and the PMP module Address line 2), a new initialization routine can be written in just a few lines:

```
#DEFINE PS2CLOCK  _RG9    // CN11 input pin
#define PS2DAT    _RG12   // any available 5V tolerant input

void initKBD( void)
{   // PS/2 keyboard
    CNEN1 = 0x0800;       // enable CN11 input change notification
    _CNIF = 0;            // clear the interrupt flag
    _CNIE = 1;            // enable the interrupt on change notification
} // initKBD
```

As per the interrupt service routine, we can use exactly the same state machine used in the previous example, adding only a couple of lines of code to make sure that we are looking at a falling edge of the clock line.

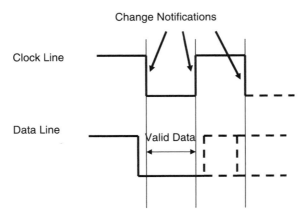

Figure 11-13. PS/2 Interface Bit Timing, Change Notification events detail.

177

In fact, when using the Input Capture module, we could choose to receive an interrupt only on the desired clock edge, while the Change Notification module will generate an interrupt both on falling and rising edges. A simple check of the status of the Clock line immediately after entering the interrupt service routine will help us tell the two edges apart:

```c
void _ISR _CNInterrupt( void)
{ // change notification interrupt service routine

// make sure it was a falling edge
    if ( PS2CLK == 0)
{
// PS/2 receiving state machine
  switch( PS2State){
        default:
        case PS2START:
            if ( ! PS2DAT)
            {
                KCount = 8;              // init bit counter
                KParity = 0;            // init parity check
                PS2State = PS2BIT;
            }
            break;

        case PS2BIT:
            KBDBuf >>=1;                 // shift in data bit
            if ( PS2DAT)
                KBDBuf += 0x80;
            KParity ^= KBDBuf;          // update parity
            if ( --KCount == 0)         // if all bit read, move on
                PS2State = PS2PARITY;
            break;

        case PS2PARITY:
            if ( PS2DAT)
                KParity ^= 0x80;
            if ( KParity & 0x80)        // if parity is odd, continue
                PS2State = PS2STOP;
            else
                PS2State = PS2START;
            break;
```

```
            case PS2STOP:
                KBDBuf >>=1;                  // shift in data bit
                if ( PS2DAT)
                    KBDBuf += 0x80;
                KParity ^= KBDBuf;            // update parity
                if ( --KCount == 0)          // if all bit read, move on
                    PS2State = PS2PARITY;
                break;
    } // switch state machine


        } // if falling edge


    // clear interrupt flag
        _CNIF = 0;


    } // CN Interrupt
```

Add the constants and variables declarations already used in the previous example:

```
#include <p24fj128ga010.h>
#include "PS2CN.h"

#define PS2DAT  _RG12        // PS2 Data input pin
#define PS2CLK  _RG9         // PS2 Clock input pin

// definition of the keyboard PS/2 state machine
#define PS2START    0
#define PS2BIT      1
#define PS2PARITY   2
#define PS2STOP     3

// PS2 KBD state machine and buffer
int PS2State;
unsigned char KBDBuf;
int KCount, KParity;

// mailbox
volatile int KBDReady;
volatile unsigned char KBDCode;
```

Package it all together in a file that we will call "PS2CN.c".

The include file "PS2CN.h" is going to be almost identical to the previous example since we are going to offer the same interface:

```
/*
**
**   PS2CN.h
**
**   PS/2 keyboard input module using Change Notification
*/

extern volatile int KBDReady;
extern volatile unsigned char KBDCode;

void initKBD( void);
```

Create a new project called "PS2CN" and add both the `.c` and the `.h` files to the project.

Finally, create a main module to test this new technique. One more time, it is going to be almost identical to the previous project:

```
/*
** PS2 KBD Test
**
*/

#include <p24fj128ga010.h>

#include "PS2CN.h"

main()
{
    TRISA = 0xff00;
    initKBD();                    // call the initialization routine

    while ( 1)
    {
        if ( KBDReady)            // wait for the flag
        {
            PORTA = KBDCode;      // fetch the key code and publish on PORTA
            KBDReady = 0;         // clear the flag
        }
    } // main loop
} //main
```

Save the project and then build the project ("Project→BuildAll") to compile and link all the modules.

To test the change notification technique, we will once more use MPLAB SIM stimulus-generation capabilities and we will repeat most of the steps performed in the previous project. Starting with the SCL Generator ("Debugger→SCLGenerator"), we will create a new Workbook. Inside the generator window, create two columns; one will be for the same PS2 Data line connected to RG12, but the other one this time will be for the PS2 Clock line connected to the CN11 Change Notification module input.

Add the same sequence of times and events to the table as used in the previous example, replacing the IC1 input column with the CN11 column. Save the Workbook as "PS2CN.sbs" and then hit the "Generate SCL" file to produce the output stimulus script file: "PS2CN.scl". Finally, activate the Stimulus Controller ("Debugger→StimulusController") and create a new Scenario. From the Stimulus Controller window, click on the Attach button and select the "PS2CN.scl" file to activate the input simulation. Save the Scenario if you want, but don't close the Controller window (you can minimize it though).

We are ready now to execute the code and test (in simulation) the proper functioning of the new PS/2 interface. Open the Watch window and add PORTA. Then set a breakpoint inside the main loop, right after the key code is copied into the PORTA register. Finally, perform a reset ("Debugger→Reset") and verify that the first event is triggered (setting both PS/2 input lines high at time 0 µs). Execute the code ("Debugger→RUN") and, if all goes well, you will see the processor stop at the breakpoint after less than a second and you will see the contents of PORTA updated to reflect the key code 0x79. Success!

Evaluating cost

Changing from the input capture to the Change Notification method was almost too easy. The two peripherals are extremely potent and, although designed for different purposes, when applied to the task at hand they performed almost identically. In the embedded world, though, you should constantly ask yourself if you could solve the problem with fewer resources, even when apparently, as in this case, there seems to be abundance. Let's evaluate the real cost of each solution by counting the resources used and their relative scarcity. When using the input capture, we are in fact using one of five IC modules available in the PIC24FJ128GA010 model. This peripheral is designed to operate in conjunction with a timer (Timer2 or Timer3), although we are not using the timing information in our application, but only the interrupt mechanism associated with the input edge trigger. When using the change notification, we are using only one of 22 possible inputs, but we are also taking control of the sole interrupt vector available to this peripheral. In other words, should we need any other input pin to be controlled by the Change Notification peripheral, we will have to share the interrupt vector, adding latency and complexity to the solution. I would call this a tie.

A third method – I/O polling

There is one more method that we could explore to interface to a PS/2 keyboard. It is the most basic one and it implies the use of a timer, set for a periodic interrupt, and its inputs can be any (5V tolerant) I/O pin of the microcontroller. In a way, this method is the most flexible from a configuration and layout point of view. It is also the most generic, as any microcontroller model, even the smallest and most inexpensive, will offer at least one timer module suitable for our purpose. The theory of operation is pretty simple. At regular intervals, an interrupt will be generated, set by the value of the period register associated with the chosen timer.

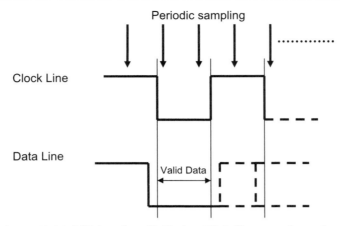

Figure 11-14. PS/2 Interface Bit Timing, I/O Polling sampling points.

We will use Timer4 this time, since we've never used it before. Hence, PR4 will be the period register. The interrupt service routine (T4Interrupt) will sample the status of the PS/2 Clock line and it will determine whether a falling edge has occurred on the PS/2 Clock line over the previous period. When a falling edge is detected, the Data line status will be considered to have received the key code. In order to determine how frequently we should perform the sampling, and therefore identify the optimal value of the PR4 register, we should look at the shortest amount of time allowed between two edges on the PS/2 clock line. This is determined by the maximum bit-rate specified for the PS/2 interface, which, according to the documentation in our possession, corresponds to about 16 kbit/s. At that rate, the clock signal can be represented by a square wave with approximately 50% duty cycle, and a period of approximately 62.5 μs. In other words, the clock line will stay low for little more than 30 μs each time a data bit is presented on the Data line, and will stay high for approximately the same amount of time, during which the next bit will be shifted out. By setting PR4 to a value that will make the interrupt period shorter than 30 μs (say 25 μs), we can guarantee that the clock line will be sampled at least once between two consecutive edges. The keyboard transmission bit rate, though, could be as slow as 10 kbit/s, giving a maximum distance between edges of about 50 μs. In that case, we would be sampling the clock and data lines twice and possibly up to three times between each clock edge. In other words, we will have to build a new state machine to detect the actual occurrence of a falling edge and to properly keep track of the PS/2 clock signal.

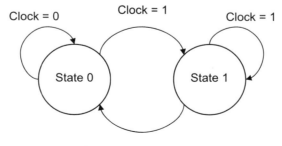

Figure 11-15. Clock-polling state machine graph.

The state machine requires only two states, and all the transitions can be summarized in Table 11-4.

State	Conditions	Effect
State0	Clock = 0	Remain in State0
	Clock = 1	Rising Edge, Transition to State1
State1	Clock = 1	Remain in State1
	Clock = 0	Falling edge detected
		Execute the Data state machine
		Transition to State0

Table 11-4. Clock-polling state machine transitions table.

When a falling edge is detected, we can still use the same state machine developed in the previous projects to read the data line. It is important to notice that in this case the value of the data line is not guaranteed to be sampled right after the actual falling edge of the clock line has occurred, but could be considerably delayed. To avoid the possibility of reading the data line outside the valid period, it is imperative to sample simultaneously both the clock and the data line. By definition (PS/2 specifications), if the clock line is low, the data can be considered valid. In practice the requirement translates into the necessity to assign both the data and clock inputs to pins of the same port. In our example we will choose to use RG12 (again) for the clock line, and RG15 for the data line. In this way, copying PORTG contents into a temporary variable, as soon as we enter the interrupt service routine, will give us an atomic action and perfect simultaneity in sampling the two lines. Here is the simplest implementation of the Clock state machine illustrated in Figure 11-15:

```
#define PS2DAT  _RG12       // PS2 Data input pin
#define PS2CLK  _RG15       // PS2 Clock input pin
#define CLKMASK 0x8000      // mask to detect the clock line
#define DATMASK 0x1000      // mask to detect the data line


unsigned char KBDBuf;
int KState;


// mailbox
volatile int KBDReady;
volatile unsigned char KBDCode;

void _ISR _T4Interrupt( void)
{
    int PS2IN;

    // sample the inputs, clock and data, at the same time
    PS2IN = PORTG;
```

```
                // Keyboard clock state machine
                if ( KState)
                {    // previous time clock was high, State1
                     if ( !(PS2IN & CLKMASK))        // PS2CLK = 0

                {// falling edge detected
                      KState = 0;                    // transition to State0

                      <<<... Insert Data state machine here!>>>


                     } // falling edge

                     else
                     { // clock still high, remain in State1

                     } // clock still high
                } // State 1

                else
                { // State 0
                     if ( PS2IN & CLKMASK)           // PS2CLK = 1
                     { // rising edge detected
                        KState = 1;         // transition to State1

                     } // rising edge

                     else
                     { // clocl still low, remain in State0

                     } // clock still low
                } // State 0

                // clear the interrupt flag
                _T4IF = 0;

     } // T4 Interrupt
```

Thanks to the periodic nature of the polling mechanism we just developed, we can add a new feature to the PS2 interface to make it more robust with minimal effort. First of all we can add a counter of the idle loops of both states of the clock state machine. This way we will be able to create a timeout, so as to be able to detect and correct error conditions should the PS/2 keyboard be disconnected during a transmission or should the receive routine lose synchronization for any reason.

The new transition table is quickly updated to include the timeout counter Ktimer.

State	Conditions	Effect
State0	Clock = 0	Remain in State0 Decrement Ktimer If Ktimer = 0, error Reset the data state machine
	Clock = 1	Rising Edge, Transition to State1
State1	Clock = 1	Remain in State1 Decrement Ktimer If Ktimer = 0, error Reset the data state machine
	Clock = 0	Falling edge detected Execute the Data state machine Transition to State0 Restart Ktimer

Table 11-5. Clock-polling (with timeout) state machine transition table.

The new transition table adds only a few instructions to our interrupt service routine.

```
void _ISR _T4Interrupt( void)
{
    int PS2IN;

    // sample the inputs, clock and data, at the same time
    PS2IN = PORTG;

    // Keyboard clock state machine
    if ( KState)
    {   // previous time clock was high, State1
        if ( !(PS2IN & CLKMASK))        // PS2CLK = 0
        {// falling edge detected
            KState = 0;                 // transition to State0
            KTimer = KMAX;              // restart the counter

            <<<... Insert Data state machine here!>>>


        } // falling edge

        else
        { // clock still high, remain in State1
            KTimer--;
            if ( KTimer ==0)            // timeout!
                PS2State = PS2START;    // reset the data state machine
```

185

```
        } // clock still high
    } // State 1

    else
    { // State 0
        if ( PS2IN & CLKMASK)                    // PS2CLK = 1
        { // rising edge detected
            KState = 1;                  // transition to State1

        } // rising edge

        else
        { // clocl still low, remain in State0
            KTimer--;
            if ( KTimer == 0)            // timeout!
                PS2State = PS2START;     // reset the data state machine
        } // clock still low
    } // State 0

    // clear the interrupt flag
    _T4IF = 0;

} // T4 Interrupt
```

Testing the I/O polling method

Let's now insert the Data state machine from the previous projects, modified to operate on the value sampled in PS2IN at the interrupt service routine entry:

```
switch( PS2State){
default:
case PS2START:
    if ( !(PS2IN & DATMASK))
    {
        KCount = 8;           // init bit counter
        KParity = 0;          // init parity check
        PS2State = PS2BIT;
    }
    break;

case PS2BIT:
    KBDBuf >>=1;                         // shift in data bit
    if ( PS2IN & DATMASK)                //PS2DAT
        KBDBuf += 0x80;
    KParity ^= KBDBuf;                   // calculate parity
    if ( --KCount == 0)                  // if all bit read, move on
        PS2State = PS2PARITY;
    break;
```

```
            case PS2PARITY:
                if ( PS2IN & DATMASK)
                    KParity ^= 0x80;
                if ( KParity & 0x80)            // if parity is odd, continue
                    PS2State = PS2STOP;
                else
                    PS2State = PS2START;
                break;

            case PS2STOP:
                if ( PS2IN & DATMASK)           // verify stop bit
                {
                    KBDCode = KBDBuf;    // write in the buffer
                     KBDReady = 1;               // set flag
                }
                PS2State = PS2START;
                break;

        } // switch
```

Let's complete this third module with a proper initialization routine.

```
void initKBD( void)
{
    // init I/Os
    _TRISG15 = 1;   // make RG15 an input pin, PS/2 Clock
    _TRISG12 = 1;   // make RG12 an input pin, PS/2 Data

    // clear the flag
    KBDReady = 0;

    PR4 = 25 * 16;  // 25 us, set the period register
    T4CON = 0x8000; // T4 on, prescaler 1:1
    _T4IF = 0;      // clear interrupt flag
    _T4IE = 1;      // enable interrupt


} // init KBD
```

This is straightforward.

Let's save it all in a module we can call "PS2T4.c". Let's create a new include file too:

```
    /*
    **
    **  PS2T4.h
    **
    **  PS/2 keyboard input library using T4 polling
    */
```

```
extern volatile int KBDReady;
extern volatile unsigned char KBDCode;


void initKBD( void);
```

It is practically identical to all previous modules' include files, and the main module will not be much different either:

```
/*
** PS2 KBD Test
**
**
*/


#include <p24fj128ga010.h>

#include "PS2T4.h"

main()
{
    TRISA = 0xff00;
    initKBD();                  // call the initialization routine

    while ( 1)
    {
        if ( KBDReady)          // wait for the flag
        {
            PORTA = KBDCode;    // fetch the key code and publish on PORTA
            KBDReady = 0;               // clear the flag
        }
    } // main loop
} //main
```

Create a new project "PS2T4" and add all three files to it. Build all and follow the same series of steps used in the previous two examples to generate a stimulus script file "PS2T4.scl". Remember that, this time, the stimulus for the clock line must be provided on the RG15 pin. Open a new scenario with the stimulus controller and attach the stimulus script file to start the simulation (remember to leave the Stimulus Controller window open in the background). Open the Watch window and add PORTA. Finally, set a breakpoint to the line after the assignment to PORTA and execute. If all goes well, even this time you should be able to see PORTA updated in the Watch window and showing a new value of 0x79. Success again!

Cost and efficiency of the solution

Comparing the cost of this solution to the previous two, we realize that the I/O polling approach is the one that gives us the most freedom in choosing the input pins and uses only one resource, a timer, and one interrupt vector. The periodic interrupt can also be seamlessly shared with other tasks to form a common time base if they all can be reduced to multiples of the polling period. The timeout feature is an extra bonus; in order to implement it in the previous techniques, we would have had to use a separate timer and another interrupt service routine in addition to the Input Capture or Change Notification modules and interrupts. Looking at the code efficiency, the Input Capture and the Change Notification methods appear to have an advantage, as an interrupt is generated only when an edge is detected. Actually, as we have seen, the input capture is the best method from this point of view, as we can select precisely the one type of edge we are interested in—that is, the falling edge of the clock line. The I/O polling method appears to require the longest interrupt routine, but the number of lines does not reflect the actual weight of the interrupt service routine. In fact, of the two nested state machines that comprise the I/O polling interrupt service routine, only a few instructions are executed at every call, resulting in a very short execution time and minimal overhead.

To verify the actual software overhead imposed by the interrupt service routines, we can perform one simple test on each one of the three implementations of the PS/2 interface. I will use the last one as an example. We can allocate one of the I/O pins (one of the LED outputs would be a logical choice) to help us visualize when the microcontroller is inside an interrupt service routine. We can set the pin on entry and reset it right before exit:

```
void _ISR _T4Interrupt( void)
{
    _RA0 = 1;                    // flag up, inside the ISR
    ...

    <<< Interrupt service routine here >>

    _RA0 = 0;                    // flag down, back to the main
}
```

Using the MPLAB SIM simulator Logic Analyzer view, we can visualize it on our computer screen. Follow the Logic Analyzer checklist so you will remember to enable the Trace buffer, and set the correct simulation speed. Select the RA0 channel and rebuild the project. To test the first two methods you will need to activate the Stimulus Controller again to simulate the inputs; without them, there are going to be no interrupts at all.

To test the polling routine, you don't need stimuli. The timer interrupt keeps coming anyway and we are particularly interested in seeing how much time is wasted by the continuous polling when no keyboard input is provided.

Let MPLAB SIM execute for a few seconds; then stop the simulation and switch back to the Logic Analyzer window. You will have to zoom in quite a bit to get an accurate picture.

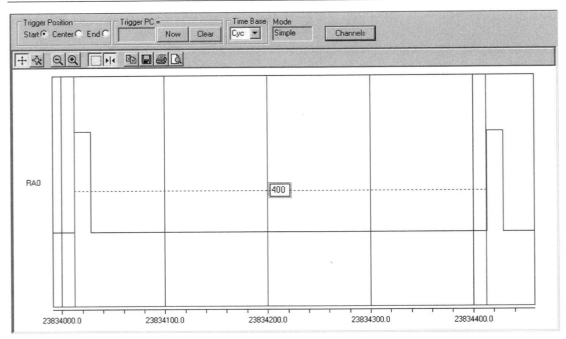

Figure 11-16. Logic Analyzer view, measuring the I/O polling period.

Activate the cursors ▶◀ and drag them to measure the number of cycles between two consecutive rising edges of RA0, marking the entry in the interrupt service routine. Since we selected a 25-µs period, you should read 400 cycles between calls (25 µs * 16 cycles/µs @32 MHz). Measuring the number of cycles between a rising edge and a falling edge of RA0 will tell us, with good approximation, how much time we are spending inside the interrupt service routine; 16 cycles is what I found. The ratio between the two quantities will give us an indication of the computing power absorbed by the PS/2 interface. In our case, that turns out to be just 2.5%.

Completing the interface: adding a FIFO buffer

Independently from the solution you will choose out of the three we explored so far, there are a few more details we need to take care of before we can claim to have completed the interface to the PS/2 keyboard. First of all, we need to add a FIFO buffering mechanism between the PS/2 interface routines and the "consumer" or the main application. So far, in fact, we have provided only a simple mailbox mechanism that can store only the last key code received. If you investigate further how the PS/2 keyboard protocol works, you will discover that when a single key is pressed and released, a minimum of three (and a maximum of five) key codes are sent to the host. If you consider shift, control and Alt-key combinations, things get a little more complicated and you realize immediately that the single-byte mailbox is not going to be sufficient. My suggestion, in fact, is to add at least a 16-byte FIFO buffer. The input to the buffer can be easily integrated with the receiver interrupt service routines so that, when a new key code is received, it is immediately inserted in the FIFO. The buffer can be declared as an array of characters and two pointers will keep track of the head and tail of the buffer in a circular scheme.

190

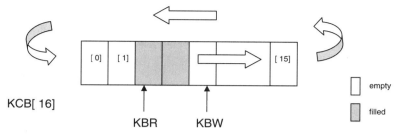

Figure 11-17. Circular buffer FIFO.

```
// circular buffer
unsigned char KCB[ KB_SIZE];

// head and tail or write and read pointers
volatile int KBR, KBW;
```

By following a few simple rules we can keep track of the buffer content:

- the write pointer KBW (or head), marks the first empty location that will receive the next key code.

- the read pointer KBR (or tail), marks the first filled location.

- when the buffer is empty, KBR and KBW are pointing at the same location.

- when the buffer is full, KBW points to the location before KBR.

- after reading or writing a character to/from the buffer, the corresponding pointer is incremented.

- upon reaching the end of the array, each pointer will wrap around to the first element of the array

Insert the following snippet of code into the initialization routine:

```
// init the circular buffer pointers
KBR = 0;
KBW = 0;
```

Then update the interrupt routine state machine STOP state:

```
case PS2STOP:
    if ( PS2IN & DATMASK)        // verify stop bit
    {
        KCB[ KBW] = KBDBuf;      // write in the buffer
        if ( (KBW+1)%KB_SIZE != KBR) // check if buffer full
            KBW++;               // else increment buffer
        KBW %= KB_SIZE;          // wrap around
    }
    PS2State = PS2START;
    break;
```

Notice the use of the "%" operator to give us the remainder of the division by the buffer size. This allows us to keep the pointers wrapping around the circular buffer.

A few considerations are required for fetching key codes from the FIFO buffer. In particular, if we choose the Input Capture or the Change Notification methods, we will need to make a new function available (getKeyCode()) to replace the mailbox/flag mechanism. The function will return FALSE if there are no key codes available in the buffer and TRUE if there is at least one key code in the buffer, and the code is returned via a pointer:

```
int getKeyCode( char *c)
{
    if ( KBR == KBW)              // buffer empty
    return FALSE;

    // buffer contains at least one key code
    *c = KCB[ KBR++];            // extract the first key code
    KBR %= KB_SIZE;                      // wrap around the pointer

    return TRUE;
} // getKeyCode
```

Notice that the extraction routine modifies only the read pointer; therefore, it is safe to perform this operation when the interrupts are enabled. Should an interrupt occur during the extraction, there are two possible scenarios:

* the buffer was empty: a new key code will be added, but the getKeyCode routine will "notice" the available character only at the next call.

* the buffer was not empty: the interrupt routine will add a new character to the buffer tail, if there is enough room.

In both cases, there are no particular concerns of conflicts or dangerous consequences.

If we choose the polling technique, there is one more option we might want to explore. In fact, since the timer interrupt is constantly active, we can use it to perform one more task for us. The idea is to maintain the simple mailbox-and-flag mechanism for delivering key codes as the interface to the receive routine, and have the interrupt constantly checking the mailbox, ready to replenish it with the content from the FIFO. This way we can confine the entire FIFO management to the interrupt service routine, making the buffering completely transparent and maintaining the simplicity of the mailbox delivery interface. The new and complete interrupt service routine for the polling I/O mechanism follows:

```
void _ISR _T4Interrupt( void)
{
    int PS2IN;
    // check if buffer available
    if ( !KBDReady && ( KBR!=KBW))
    {
        KBDCode = KCB[ KBR++];
        KBR %= KB_SIZE;
        KBDReady = 1;             // signal character available
    }

    // sample the inputs clock and data at the same time
    PS2IN = PORTG;
```

```
        // Keyboard state machine
        if ( KState)
        {   // previous time clock was high KState 1
            if ( !(PS2IN & CLKMASK))    // PS2CLK = 0
            {   // falling edge detected,
                KState = 0;                 // transition to State0
                KTimer = KMAX;              // restart the counter

                switch( PS2State){
                default:
                case PS2START:
                    if ( !(PS2IN & DATMASK))
                    {
                        KCount = 8;         // init bit counter
                        KParity = 0;        // init parity check
                        PS2State = PS2BIT;
                    }
                    break;

                case PS2BIT:
                    KBDBuf >>=1;                // shift in data bit
                    if ( PS2IN & DATMASK)       //PS2DAT
                        KBDBuf += 0x80;
                    KParity ^= KBDBuf;          // calculate parity
                    if ( --KCount == 0)         // if all bit read, move on
                        PS2State = PS2PARITY;
                    break;

                case PS2PARITY:
                    if ( PS2IN & DATMASK)
                        KParity ^= 0x80;
                    if ( KParity & 0x80)        // if parity is odd, continue
                        PS2State = PS2STOP;
                    else
                        PS2State = PS2START;
                    break;

                case PS2STOP:
                    if ( PS2IN & DATMASK)       // verify stop bit
                    {
                        KCB[ KBW] = KBDBuf;     // write in the buffer
                        if ( (KBW+1)%KB_SIZE != KBR) // check if buffer full
                            KBW++;              // else increment buffer
                        KBW %= KB_SIZE;         // wrap around
                    }
                    PS2State = PS2START;
                    break;
```

```
            } // switch
        } // falling edge
        else
        { // clock still high, remain in State1
            KTimer--;
            if ( KTimer ==0)
                PS2State = PS2START;
        } // clock still high
    } // Kstate 1
    else
    { // Kstate 0
        if ( PS2IN & CLKMASK)      // PS2CLK = 1
        { // rising edge, transition to State1
            KState = 1;
        } // rising edge
        else
        { // clocl still low, remain in State0
            KTimer--;
            if ( KTimer == 0)
                PS2State = PS2START;
        } // clock still low
    } // Kstate 0

    // clear the interrupt flag
    _T4IF = 0;

} // T4 Interrupt
```

Completing the interface: performing key codes decoding

So far we have been talking exclusively about key codes and you might have assumed that they match the ASCII codes for each key. For example, if you press the "A" key on the keyboard, you would expect the corresponding ASCII code (0x41) to be sent. Unfortunately, this is not the case. For historical reasons, even the newest USB keyboards are still bound to use "scan codes" where each key is assigned a numerical value that is related to the original implementation of the keyboard-scanning firmware (which used an 8048 microcontroller) for the first IBM PC keyboard circa 1980. The fact that the translation from key codes to a specific character set happens at a higher level (performed by Windows keyboard drivers) is actually a good thing, since it provides a generic mechanism to support many different international keyboard layouts. Keep in mind also that, for historical reasons, there are at least three different and partially compatible "scan code sets." Fortunately, by default, all keyboards support scan code set #2, which is the one we will focus on in the following section.

Each time a key is pressed (any key, including a shift or control key), the scan code associated to it is sent to the host. This is called the "make code." But also, as soon as the same key is released, a new sequence of scan codes is sent to the host. This is called the "break code." The break code is typically composed of the same scan code but prefixed with the code "0xF0". Some keys can have a two-byte-long make code (typically the Ctrl, Alt and arrows) and consequently the break code is three bytes long.

Key	Make Code	Break Code
"A"	1C	F0,1C
"5"	2E	F0,2E
"F10"	09	F0,09
Right Arrow	E0, 74	E0, F0, 74
Right "Ctrl"	E0, 14	E0, F0, 14

Table 11-6. Example of make and break codes used in Scan Code Set 2 (default).

In order to process this information and translate the scan codes intro proper ASCII, we will need a table that will help us map the basic scan codes for a basic U.S. English keyboard layout.

```
// PS2 keyboard codes (standard set #2)
const char keyCodes[128]={
            0,   F9,    0,   F5,   F3,   F1,   F2, F12,   //00
            0,  F10,   F8,   F6,   F4,  TAB,  '`',    0,   //08
            0,    0,L_SHFT,  0,L_CTRL,'q','1',    0,   //10
            0,  'z',  's',  'a',  'w',  '2',    0,   //18
            0,  'c',  'x',  'd',  'e',  '4',  '3',    0,   //20
            0,  ' ',  'v',  'f',  't',  'r',  '5',    0,   //28
            0,  'n',  'b',  'h',  'g',  'y',  '6',    0,   //30
            0,    0,  'm',  'j',  'u',  '7',  '8',    0,   //38
            0,  ',',  'k',  'i',  'o',  '0',  '9',    0,   //40
            0,  '.',  '/',  'l',  ';',  'p',  '-',    0,   //48
            0,    0,'\'',    0,  '[',  '=',    0,    0,   //50
         CAPS, R_SHFT,ENTER, ']',   0,0x5c,    0,    0,   //58
            0,    0,    0,    0,    0,    0, BKSP,    0,   //60
            0,  '1',    0,  '4',  '7',    0,    0,    0,   //68
            0,  '.',  '2',  '5',  '6',  '8',  ESC,  NUM,   //70
          F11,  '+',  '3',  '-',  '*',  '9',    0,    0   //78
    };
```

Notice that the array has been declared as `const` so that it will be allocated in program memory space to save more precious RAM space.

It will also be convenient to have a similar table available for the shift function of each key.

```
const char keySCodes[128] = {
            0,   F9,    0,   F5,   F3,   F1,   F2, F12,   //00
            0,  F10,   F8,   F6,   F4,  TAB,  '~',    0,   //08
            0,    0,L_SHFT,  0,L_CTRL,'Q','!',    0,   //10
            0,    0,  'Z',  'S',  'A',  'W',  '@',    0,   //18
            0,  'C',  'X',  'D',  'E',  '$',  '#',    0,   //20
            0,  ' ',  'V',  'F',  'T',  'R',  '%',    0,   //28
            0,  'N',  'B',  'H',  'G',  'Y',  '^',    0,   //30
            0,    0,  'M',  'J',  'U',  '&',  '*',    0,   //38
            0,  '<',  'K',  'I',  'O',  ')',  '(',    0,   //40
            0,  '>',  '?',  'L',  ':',  'P',  '_',    0,   //48
            0,    0,'\"',    0,  '{',  '+',    0,    0,   //50
```

```
          CAPS, R_SHFT,ENTER,  '}',   0,  '|',   0,    0,    //58
                0,   0,    0,    0,    0,   0, BKSP,   0,    //60
                0,  '1',   0,  '4',  '7',   0,    0,    0,    //68
                0,  '.',  '2',  '5',  '6',  '8', ESC, NUM,    //70
              F11,  '+',  '3',  '-',  '*',  '9',   0,    0    //78
          };
```

For all the ASCII characters, the translation is straightforward, but we will have to assign special values to the function, shift and control keys. Only a few of them will find a corresponding code in the ASCII set:

```
// special function characters
#define TAB      0x9
#define BKSP     0x8
#define ENTER    0xd
#define ESC      0x1b
```

For all the others, we will have to create our own conventions, or, until we have a use for them, we might just ignore them and assign them a common code (0):

```
#define L_SHFT   0x12
#define R_SHFT   0x12
#define CAPS     0x58
#define L_CTRL   0x0
#define NUM      0x0
#define F1       0x0
#define F2       0x0
#define F3       0x0
#define F4       0x0
#define F5       0x0
#define F6       0x0
#define F7       0x0
#define F8       0x0
#define F9       0x0
#define F10      0x0
#define F11      0x0
#define F12      0x0
```

The following routine `getC()` performs the basic translations for the most common keys and it takes care of the shift status as well as the CAPS key toggling:

```
int CapsFlag=0;

char getC( void)
{
    unsigned char c;

    while( 1)
    {
        while( !KBDReady);        // wait for a key to be pressed
        // check if it is a break code
```

```
        while (KBDCode == 0xf0)
        {   // consume the break code
            KBDReady = 0;
            // wait for a new key code
            while ( !KBDReady);
            // check if the shift button is released
            if ( KBDCode == L_SHFT)
                CapsFlag = 0;
            // and discard it
            KBDReady = 0;
            // wait for the next key
            while ( !KBDReady);
        }
        // check for special keys
        if ( KBDCode == L_SHFT)
        {
            CapsFlag = 1;
            KBDReady = 0;
        }
        else if ( KBDCode == CAPS)
        {
            CapsFlag = !CapsFlag;
            KBDReady = 0;
        }

        else // translate into an ASCII code
        {
            if ( CapsFlag)
                c = keySCodes[KBDCode%128];
            else
                c = keyCodes[KBDCode%128];
            break;
        }
    }
    // consume the current character
    KBDReady = 0;

    return ( c);
} // getC
```

Post-flight briefing

In this lesson we have learned how to interface to a PS/2 computer keyboard, exploring three alternative methods. This gave us the perfect opportunity to exercise two new peripheral modules: the Input Capture and the Change Notification modules. We also discussed methods to implement a FIFO buffer and polished our interrupt management skills. Throughout the entire lesson, our focus has been constantly on balancing the use of resources and the performance offered by each solution.

Tips and tricks

Stalling transmissions from the keyboard – Open-Drain Output Control

Each PS/2 keyboard has an internal FIFO buffer 16 key codes deep. This allows the keyboard to accumulate the user input even when the host is not ready to receive. The host, as we mentioned at the very beginning of this chapter, has the option to stall the communication by pulling the clock line low at any given point in time (for at least 100 µs) and can hold it low for the desired period of time. When the clock line is released, the keyboard will resume transmissions. It will retransmit the last key code, if it had been interrupted, and will offload its FIFO buffer.

To exercise our right to stall the keyboard transmissions as a host, we have to control the clock line with an output using an open drain driver. Fortunately, this is easy with the PIC24, thanks to its configurable I/O port modules. In fact, each I/O port (PORTx) has an associated control register (ODCx) that can individually configure each pin output driver to operate in open-drain mode.

Note: This feature is extremely useful in general to interface PIC24 outputs to any 5V device.

In our example, turning the PS/2 clock line into an open-drain output would require only a few lines of code:

```
_ODG13 = 1;       // configure the PORTG pin 13 output driver in open-drain
_LATG13 = 1;      // initially let the output in pull up
_TRISG13 = 0;     // enable the output driver
```

Note that, as usual for all PIC® microcontrollers, even if a pin is configured as an output, its current status can still be read as an input. So there is no reason to switch continuously between input and output when we alternate stalling and receiving characters from the keyboard.

Exercises

1. Add a function to send commands to the keyboard to control the status LEDs and set the key repeat rate.

2. Replace the "`stdio.h`" library input function `read()` to redirect the keyboard input from the `stdin` stream.

3. Add support for a PS/2 mouse interface.

Books

* Anderson F. (2003)

   Flying the Mountains

   McGraw-Hill, New York, NY

   Flying the mountains requires extra caution and preparation. This could be the next challenge after you have completed your private pilot license.

Links

* *http://www.computer-engineering.org/*

   This is an excellent web site where you will find a lot of useful documentation on the PS/2 keyboard and mouse interface.

CHAPTER **12**

The Dark Screen

I have always liked driving the car at night. Generally there is less traffic, the air is always cooler and, unless I am really tired, the lights of the vehicles in the other direction never really bother me much. But when my instructor proposed a first cross-country flight at night, I got a little worried. The idea of staring at a windshield filled with pitch black void…was a little frightening, I have to admit. However, the actual experience a week later converted me forever. Sure, night flying is a bit more serious stuff than the usual around-the-pattern practice. There is more careful planning involved, but it is just so rewarding. Flying over an uninhabited area fills the screen with so many stars that a city boy like me has hardly ever seen—it feels like flying a starship to another solar system. Flying over or near a large city transforms the grey and uniform spread of concrete of alternating parking lots and housing developments into a wonderful show of lights—it's like Christmas as far as the eye can see. Turns out, the screen is never really dark. It's a big show and it is on every night.

Flight plan

In this lesson we will consider techniques to interface to a TV screen or, for that matter, to any display that can accept a standard composite video signal. It will be a good excuse to use new features of several peripheral modules of the PIC24 and review new programming techniques. Our first project objective will be to get a nice dark screen (a well-synchronized video frame), but we will soon fill it up with several entertaining graphical applications.

The flight

There are many different formats and standards today in use in the world of video, but perhaps the oldest and most common one is the so-called "composite" video format. This is what was originally used by the very first TV sets to appear in the consumer market, and today it represents the minimum common denominator of every video display, whether a modern high-definition flat-screen TV of the latest generation, a DVD player, or a VHS tape recorder. All video devices are based on the same basic concept: the image is "painted" one line at a time, starting from the top left corner of the screen and moving horizontally to the right edge, then quickly jumping back to the left edge at a lower position and painting a second line, and so on and on, in a zig-zag motion, until the entire screen has been scanned. Then the process repeats and the entire image is refreshed fast enough for our eyes to be tricked into believing that the entire image is present at the same time, and if there is motion, it is fluid and continuous.

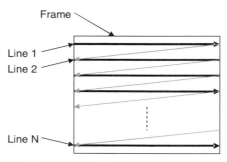

Figure 12-1. Video image scanning.

In different parts of the world, slightly incompatible systems have been developed over the years, but the basic mechanism remains the same. What changes eventually is the number of lines composing the image, the refreshing frequency, and the way the color information is encoded.

	US	**Europe, Asia**	**France and others**
Standard	NTSC	PAL	SECAM
Frames per second	29.97*	25	25
Number of lines	525	625	625

* NTSC used to be 30 frames per second, but the introduction of the new color standard changed it to 29.97, to accommodate for a specific frequency used by the "color subcarrier" crystal oscillator.

Table 12-1. International video standard examples.

Table 12-1 illustrates three of the most commonly used video standards adopted in the US, Europe and Asia. All those standards encode the "luminance" information (that is, the underlying black-and-white image) together with synchronization information in a similarly defined composite signal.

The name "composite" is used to describe the fact that three different pieces of information are combined into one video signal: the actual luminance signal and both horizontal and vertical synchronization information.

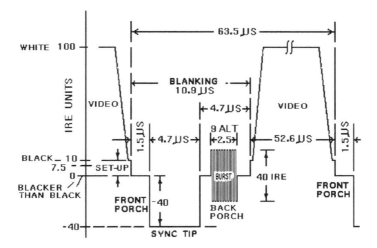

Figure 12-2. NTSC composite signal, horizontal line detail.

The horizontal line signal is in fact composed of:

1. the horizontal synchronization pulse, used by the display to identify the beginning of each line.

2. the so-called back porch, that creates a dark frame around the image.

3. the actual line luminosity signal; the higher the voltage, the more luminous the point.

4. the so-called front porch, producing the right edge of the image.

The color information is transmitted separately, modulated on a high frequency subcarrier. The three main standards differ significantly in the way they encode the color information but, for our purposes, it will be easy to ignore the problem altogether to obtain a simple black-and-white display output.

All these standard systems utilize a technique called "interlacing" to provide a (relatively) high-resolution output while requiring a reduced bandwidth. In practice, only half the number of lines is transmitted and painted on the screen in each frame. Alternate frames present only the odd or the even lines composing the picture so that the entire image content is effectively updated only at half the refresh rate (25 Hz and 30 Hz, respectively for PAL and NTSC). This is effective for typical TV broadcasting but can produce an annoying flicker when text and especially horizontal lines are displayed, as is often the case in computer monitor applications. For this reason all modern computer displays do not use "interlaced" but progressive scanning. Most modern TV sets, and especially those using LCD and plasma technologies, perform a deinterlacing of the received broadcast image. In our project we will avoid "interlacing" as well, sacrificing half of the image resolution in favor of a more stable and readable display output. In other words, we will transmit frames of 262 lines (for NTSC) at the double rate of 60 frames per second. Readers that have easier access to PAL or SECAM TV sets/monitors will find it relatively easy to modify the project for 312-line resolution with a refresh rate of 50 frames per second.

A complete video frame signal is represented in Figure 12-3.

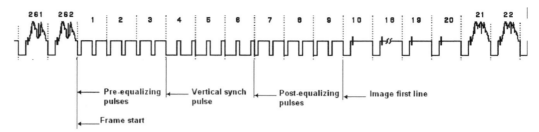

Figure 12-3. A complete video frame signal.

Notice how, out of the total number of lines composing each frame, three line periods are filled by prolonged synchronization pulses to provide the vertical synchronization information, identifying the beginning of each new frame. They are preceded and followed by groups of three additional lines, referred to as the pre- and post-equalization lines.

Generating the composite video signal

If we limit the scope of the project to generating a simple black-and-white image (no gray shades, no color) and a noninterlaced image as well, we can simplify the hardware and software requirements of our project considerably. In particular, the hardware interface can be reduced to just three resistors of appropriate value connected to two digital I/O pins. One of the I/O pins will generate the synchronization pulses and the other I/O pin will produce the actual luminance signal.

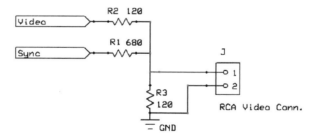

Figure 12-4. Simple hardware interface for NTSC video output.

The values of the three resistors must be selected so that the relative amplitudes of the luminance and synchronization signals are close to the standard NTSC specifications, the signal total amplitude is close to 1V peak to peak, and the output impedance of the circuit is approximately 75 ohms. With the standard resistor values shown in the previous picture, we can satisfy such requirements and generate the three basic signal levels required to produce a black-and-white image:

Signal Feature	Sync	Video
Sync pulse	0	0
Black level	1	0
White level	1	1

Table 12-2. Generating Luminance and Synchronization pulses.

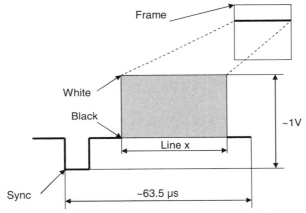

Figure 12-5. Simplified NTSC composite signal.

Since we are not going to utilize the interlacing feature, we can also simplify the pre-equalization, vertical synchronization and post-equalization pulses by producing a single horizontal synchronization pulse per each period, as illustrated in Figure 12-6.

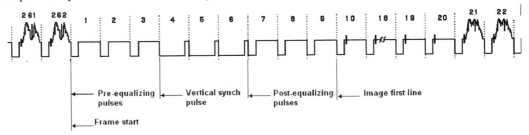

Figure 12-6. Simplified NTSC video frame (noninterlaced).

The problem of generating a complete video output signal can now be reduced to (once more) a simple state machine that can be driven by a fixed period time base produced by a single timer interrupt. The state machine will be quite trivial, as each state will be associated to one type of line composing the frame, and it will repeat for a fixed number of times before transitioning to the next state.

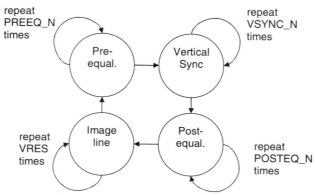

Figure 12-7. Video state machine graph.

A simple table will help describe the transitions from each state:

State	Repeat	Transition to
Pre-equal	PREEQ_N times	Vertical Sync
Vertical Sync	3 times	Post-equal
Post-equal	POSTEQ_N times	Image line
Image line	VRES times	Pre-equal

Table 12-3. Video state machine transitions table.

While the number of vertical synchronization lines is fixed and prescribed by the NTSC video standard, the number of lines effectively comprising the image inside each frame is up to us to define (within limits, of course). Although in theory we could use all of the lines available to display the largest possible amount of data on the screen, we will have to consider some practical limitations, in particular the RAM memory available to store the video image inside the PIC24FJ128GA010 microcontroller. These limitations will dictate a specific number (VRES) of lines to be used for the image while all the remaining (up to the NTSC standard line count) will be left blank.

In practice, if V_NTSC is the total number of lines composing a standard NTSC video frame and VRES is the desired vertical resolution, we will determine a value for PREEQ_N and POSTEQ_N as follows:

```
#define V_NTSC    262     // total number of lines composing a frame
#define VSYNC_N   3       // V sync lines

// count the number of remaining black lines top+bottom
#define VBLANK_N    (V_NTSC -VRES - VSYNC_N)

#define PREEQ_N    VBLANK_N /2         // pre equalization + bottom blank lines
#define POSTEQ_N   VBLANK_N - PREEQ_N  // post equalization + top blank lines
```

If we choose Timer3 to generate the time base, we can initialize its period register PR3 to produce an interrupt with the prescribed period and create an interrupt service routine where we will place the state machine. Here is a skeleton of the interrupt service routine on which we will grow the complete video generator logic.

```
// next state table
int VS[4] = { SV_SYNC, SV_POSTEQ, SV_LINE, SV_PREEQ};
// next counter table
int VC[4] = { VSYNC_N,  POSTEQ_N,    VRES,   PREEQ_N};


void _ISRFAST _T3Interrupt( void)
{
    // Start a Sync pulse
    SYNC = 0;

    // decrement the vertical counter
    VCount--;

    // vertical state machine
    switch ( VState) {
        case SV_PREEQ:
            // horizontal sync pulse
            ...
             break;

        case SV_SYNC:
            // vertical sync pulse
            ...
            break;

        case SV_POSTEQ:
            // horizontal sync pulse
            ...
            break;

        default:
        case SV_LINE:
        ...
    } //switch

    // advance the state machine
    if ( VCount == 0)
    {
        VCount = VC[ VState];
        VState = VS[ VState];
    }

    // clear the interrupt flag
    _T3IF = 0;

} // T3Interrupt
```

Once inside the interrupt service routine, we can immediately lower the Sync output pin to start generating the horizontal sync pulse, but we need a different mechanism to provide us with the right timing (approx 4.5 µs) to complete the pulse (rising edge) and produce the rest of the horizontal line waveform. There are of course several options we can explore:

1. use a short delay loop using a counter.

2. use a second timer, and associated interrupt service routine.

3. use the output compare modules and the associated interrupt service routines.

The first solution is probably the simplest to code, but has the clear disadvantage of wasting a large number of processor cycles (4.5 µs × 16 cycles per microsecond = 72 cycles), which repeated each horizontal line period (63.5 µs or about 1018 cycles) would add up to as much as 7% of the total processing power available.

The second solution is clearly more efficient, and by now we have ample experience in using timer interrupts and their interrupt service routines to execute small state machines.

The third solution involves the use of a new peripheral we have not yet explored in the previous chapters and which deserves a little more attention.

Using the Output Compare modules

The PIC24FJ128GA010 microcontroller has five Output Compare peripheral modules that can be used for a variety of applications including: single pulse generation, continuous pulse generation, and pulse width modulation (PWM). Each module can be associated to one of two 16-bit timers (Timer2 or Timer3) and has one output pin that can be configured to toggle and produce rising or falling edges if necessary. Most importantly each module has an associated and independent interrupt vector.

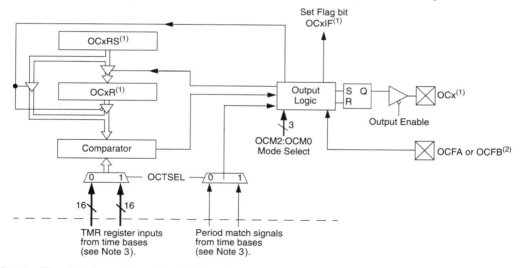

Note 1: Where 'x' is shown, reference is made to the registers associated with the respective output compare channels 1 through 8.
2: OCFA pin controls OC1-OC4 channels. OCFB pin controls OC5-OC8 channels.
3: Each output compare channel can use one of two selectable time bases. Refer to the device data sheet for the time bases associated with the module.

Figure 12-8. Output compare module block diagram.

When used in single pulse mode specifically, the OCxR register can be used to determine the instant (relative to the value of the selected timer) when the interrupt event will be triggered and, if desired, an output pin will be set/reset or toggled as required.

Upper Byte:							
U-0	U-0	R/W-0	U-0	U-0	U-0	U-0	U-0
—	—	OCSIDL	—	—	—	—	—
bit 15							bit 8

Lower Byte:							
U-0	U-0	U-0	R-0 HC	R/W-0	R/W-0	R/W-0	R/W-0
—	—	—	OCFLT	OCTSEL	OCM2	OCM1	OCM0
bit 7							bit 0

Figure 12-9. The Output Compare control register OCxCON.

The OCxCON register is the only configuration register required to control each of the output compare modules.

In our application the output compare mechanism can be quite useful as there are two precise instants where we need to take action: the end of the horizontal synchronization pulse, when generating a pre/post-equalization or a vertical synchronization line, and the end of the back porch, where the actual image begins.

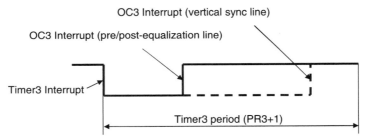

Figure 12-10. Interrupt sequence for a synchronization line.

We will choose to use one of the Output Compare modules (OC3 will be our choice) to help us identify precisely the end of the synchronization pulse. We will not need to use the associated output pin (RD2), but rather in the corresponding interrupt service routine we will raise the Sync signal.

```
void _ISRFAST _OC3Interrupt( void)
{
    SYNC = 1;   // bring the output up to the black level
    _OC3IF = 0; // clear the interrupt flag
} // OC3Interrupt
```

The OC3CON control register will be set so as to activate the output compare module in the single pulse mode (OCM=001) and to use Timer3 as the reference time base (OCTSEL=1).

We will also initialize the OC3R register with the selected timing value depending on the type of line (state of the state machine) as follows:

```
// vertical state machine
switch ( VState) {
    case SV_PREEQ:
        // horizontal sync pulse
        OC3R = HSYNC_T;
        OC3CON = 0x0009;     // single event mode
        break;

    case SV_SYNC:
        // vertical sync pulse
        OC3R = H_NTSC - HSYNC_T;
        OC3CON = 0x0009;     // single event mode
        break;

    case SV_POSTEQ:
        // horizontal sync pulse
        OC3R = HSYNC_T;
        OC3CON = 0x0009;     // single event mode
    ...
```

When generating a video line, we will use a second Output Compare module (OC4) to mark the end of the back porch and the corresponding interrupt service routine will be used to initiate the streaming of the actual image line.

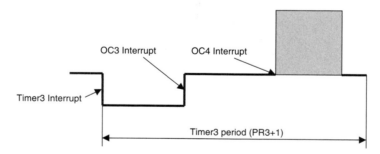

Figure 12-11. Interrupt sequence for a video line.

```
case SV_LINE:
    // activate OC3 for the end of the horizontal sync pulse
    OC3R = HSYNC_T;
    OC3CON = 0x0009;     // single event

    // activate OC4 for the end of the back porch
    OC4R = HSYNC_T + BPORCH_T;
    OC4CON = 0x0009;     // single event
    ...
    break;
```

Memory allocation

So far we have been working on the generation of the synchronization signals composing the NTSC video waveform controlled by only one of two I/Os of the simple hardware interface. The second I/O will be used once we are generating one of the lines containing the actual image. Toggling the Video I/O, we can alternate segments of the line that will be painted in white (1) or black (0). Since the NTSC standard specifies a maximum luminance signal bandwidth of 4.2 MHz, and the space between front and back porch is 52 µs wide, it follows that the maximum number of alternate segments (cycles) of black and white we can display is 218, (52 × 4.2) or in other words, our maximum theoretical horizontal resolution is 436 pixels per line (assuming the screen is completely used from side to side). The maximum vertical resolution is given by the total number of lines making up each NTSC standard frame minus the minimum number of equalization and vertical synchronization lines that gives 253. If we were to generate the largest possible image, it would be composed of an array of 253 × 436 pixels, or 110,308 pixels. Further, if one bit was used to represent each pixel that would require us to allocate an array of 13.5 kbytes, way too large to fit in the 8 kbytes available within the PIC24FJ128GA010 RAM. In practice, while it is nice to be able to generate a high-resolution output, we need to make sure that the image will fit in the available RAM memory and possibly leave enough space for an actual application to run comfortably, allowing for adequate room for the stack and application variables. While there is an almost infinite number of possible combinations of the horizontal and vertical resolution values that will give an acceptable memory size, there are two considerations that we will use to pick the perfect numbers: making the horizontal resolution a multiple of 16 will make the math involved in determining the position of a pixel in the memory map easier, assuming we use an array of integers. Also, making the two resolution values in an approximate ratio of 4:3 will avoid image geometrical distortions (in other words, circles drawn on the screen will look like circles rather then ovals).

Choosing a horizontal resolution of 256 pixels (HRES) and a vertical resolution of 192 lines (VRES) we obtain an image memory requirement of 6,144 bytes (256 × 192/8), leaving as much as 2,048 bytes for stack and application variables.

Using the C30 compiler, we can easily allocate a single array of integers (grouping 16 pixels at a time in each word) to contain the entire image memory map. But we need to make sure that the entire contents of the array are addressable and this is not possible if we declare it simply as a near variable (the default when using the small memory model). Near variables must be found within the first 8 kbytes of the data addressing space but this space also includes the special function registers area and the PSV area. The best way to avoid an allocation error message is to explicitly declare the video memory map with a `far` attribute:

```
#define _FAR __attribute__(( far))

int _FAR VMap[VRES * (HRES/16)];
```

This ensures that access to the array elements is performed via pointers, something we would have done anyway both when reading and writing to the array.

Image serialization

If each image line is represented in memory in the VMap array by a row of 16 integers, we will need to serially output each bit (pixel) in a timely fashion in the short amount of time (52 µs) between the back and the front porch part of the composite video waveform.

In other words, we will need to set or reset the chosen Video output pin with a new pixel value each 200 ns or better. This would translate into about three machine cycles between pixels, way too fast for a simple shift loop even if we plan on coding it directly in assembly. Worse, even assuming we managed to squeeze the loop in so tight, we would end up using an enormous percentage of the processing power for the video generation, leaving very few processor cycles for the main application (<18% in the best case). Fortunately, there is one peripheral of the PIC24 that can help us efficiently serialize the image data. It's the SPI synchronous serial communication module.

In a previous chapter we used the SPI2 port to communicate with a serial EEPROM memory. In that chapter we noted how the SPI module is in fact composed of a simple shift register that can be clocked by an external clock signal (when in slave mode) or by an internal clock (when in master mode). In our new project we can use the SPI1 module as a master, connecting only the SDO (serial data output) directly to the Video pin of the hardware interface, leaving the SDI (data input) unused and the SCK (clock output) and SS (slave select) pins disabled. Among the many new and advanced features of the PIC24 SPI module, two fit our video application particularly well: the ability to operate in 16-bit mode and a powerful 8-level-deep FIFO buffer. Operating in 16-bit mode, we can practically double the transfer speed of data between the image memory map and the SPI module. Enabling the 8-level-deep FIFO buffer we can load up to 128 pixels (8 words × 16 bits) at a time in the SPI buffer and quickly return from the interrupt service routine, only to return 25 μs later for a second final load, maximizing the efficiency of the video generator by requiring only two short bursts of activity for each image line.

We can now write the interrupt service routine for the second Output Compare module, configured by the state machine to be activated right after the back porch to produce the actual image line output:

```
void _ISRFAST _OC4Interrupt( void)
{
    // load SPI FIFO with 8 x 16-bit words = 128 pixels
    SPI1BUF = *VPtr++;
    SPI1BUF = *VPtr++;
    SPI1BUF = *VPtr++;
    SPI1BUF = *VPtr++;
    SPI1BUF = *VPtr++;
    SPI1BUF = *VPtr++;
    SPI1BUF = *VPtr++;
    SPI1BUF = *VPtr++;

    if ( --HCount > 0)
    { // activate again in time for the next SPI load
        OC4R += ( PIX_T * 7 * 16);
        OC4CON = 0x0009;    // single event
    }

    // clear the interrupt flag
    _OC4IF = 0;

} // OC4Interrupt
```

Notice how the interrupt service routine reconfigures the OC4 module for a second burst (the second half of the image line) after loading the first 128 pixel data in the SPI buffer.

Now that we have identified all the pieces of the puzzle, we can write the complete initialization routine for all the modules required by the video generator:

```c
void initVideo( void)
{
    // set the priority levels
      _T3IP = 4;                 // this is the default value anyway
      _OC3IP = 4;
      _OC4IP = 4;

      TMR3 = 0;              // clear the timer
      PR3 = H_NTSC; // set the period register to NTSC line

    // 2.1 configure Timer3 modules
      T3CON = 0x8000;         // enabled, prescaler 1:1, internal clock

    // 2.2 init Timer3/OC3/OC4 Interrupts, clear the flag
      _OC3IF = 0;    _OC3IE = 1;
      _OC4IF = 0; _OC4IE = 1;
      _T3IF = 0;    _T3IE = 1;

    // 2.3 init the processor priority level
      _IP = 0;                // this is the default value anyway

    // init the SPI1
    if ( PIX_T == 2)
      SPI1CON1 = 0x043B;   // Master, 16 bit, disable SCK/SS, prescale 1:3

    else
      SPI1CON1 = 0x0437;   // Master, 16 bit, disable SCK/SS, prescale 1:2

    SPI1CON2 = 0x0001;       // Enhanced mode, 8 x FIFO
    SPI1STAT = 0x8000;       // enable SPI port

    // init PORTF for the Sync
    _TRISG0 = 0;              // output the SYNC pin

    // init the vertical sync state machine
    VState = SV_PREEQ;
    VCount = PREEQ_N;

} // initVideo
```

Notice how the parameter PIX_T can be used to select different SPI clock prescaling values so as to adapt to different horizontal resolution requirements. Setting PIX_T = 3 will provide the least image distortion by giving each pixel 3 clock cycles for a total of 187.5 ns, very close to the 200-ns value previously calculated for the 256-pixel horizontal resolution.

Building the video module

We can now complete the coding of the entire video state machine, adding all the definitions and pin assignments necessary:

```
/*
** NTSC Video using T3 and Output Compare interrupts
**
*/

#include <p24fj128ga010.h>
#include "Graphic.h"

// I/O definitions
#define SYNC      _LATG0   // output
#define SDO       _RF8     // SPI1 SDO

// timing definitions for NTSC video vertical state machine
#define V_NTSC    262      // total number of lines composing a frame
#define VSYNC_N   3        // V sync lines

// count the number of remaining black lines top+bottom
#define VBLANK_N    (V_NTSC -VRES - VSYNC_N)

#define PREEQ_N    VBLANK_N /2          // pre equalization + bottom blank lines
#define POSTEQ_N   VBLANK_N - PREEQ_N   // post equalization + top blank lines

// definition of the vertical sync state machine
#define SV_PREEQ    0
#define SV_SYNC     1
#define SV_POSTEQ   2
#define SV_LINE     3

// timing definitions for NTSC video horizontal state machine
#define H_NTSC   1018    // total number of Tcy in a line (63.5us)
#define HSYNC_T  90      // Tcy in a horizontal sync pulse
#define BPORCH_T 90      // Tcy in a back porch
#define PIX_T    3       // Tcy in each pixel, valid values are only 2 or 3

#define _FAR __attribute__(( far))

int _FAR VMap[VRES * (HRES/16)];

volatile int *VPtr;
volatile int HCount, VCount, VState, HState;
```

```
// next state table
int VS[4] = { SV_SYNC, SV_POSTEQ, SV_LINE, SV_PREEQ};
// next counter table
int VC[4] = { VSYNC_N,  POSTEQ_N,    VRES,   PREEQ_N};


void _ISRFAST _T3Interrupt( void)
{
    // Start a Sync pulse
    SYNC = 0;

    // decrement the vertical counter
    VCount--;

    // vertical state machine
    switch ( VState) {
        case SV_PREEQ:
            // horizontal sync pulse
            OC3R = HSYNC_T;
            OC3CON = 0x0009;    // single event
            break;

        case SV_SYNC:
            // vertical sync pulse
            OC3R = H_NTSC - HSYNC_T;
            OC3CON = 0x0009;    // single event
            break;

        case SV_POSTEQ:
            // horizontal sync pulse
            OC3R = HSYNC_T;
            OC3CON = 0x0009;    // single event
            // on the last posteq prepare for the new frame
            if ( VCount == 0)
            {
                VPtr = VMap;
            }
            break;

        default:
        case SV_LINE:
            // horizontal sync pulse
            OC3R = HSYNC_T;
            OC3CON = 0x0009;    // single event

            // activate OC4 for the SPI loading
            OC4R = HSYNC_T + BPORCH_T;
```

```
        OC4CON = 0x0009;    // single event
        HCount = HRES/128;  // loads 8x16 bits at a time
        break;
} //switch

// advance the state machine
if ( VCount == 0)
{
    VCount = VC[ VState];
    VState = VS[ VState];
}

// clear the interrupt flag
_T3IF = 0;

} // T3Interrupt
```

To make it a complete library module we will need to add the interrupt service routines presented for the Output Compare modules OC3 and OC4 illustrated in the previous sections of this chapter, as well as a couple of additional accessory functions that will follow:

```
void clearScreen( void)
{
    int i, j;
    int *v;

    v = (int *)&VMap[0];

    // clear the screen
    for ( i=0; i < (VRES*( HRES/16)); i++)
        *v++ = 0;
} //clearScreen


void haltVideo( void)
{
    T3CONbits.TON = 0;   // turn off the vertical state machine
} //haltVideo

void synchV( void)
{
    while ( VCount != 1);
} // synchV
```

In particular, clearScreen will be useful to initialize the image memory map, the VMap array, while haltVideo will be useful to suspend the video generation should an important task/application require 100% of the PIC24 processing power.

The synchV function can be used to synchronize a task to the video generator; this function will return only when the video generator has started "painting" the last line of the screen. This can be useful for graphic applications to minimize flicker and/or provide more fluid scrolling and motion.

Save all of these functions in a file called "graphic.c" and add this file to a new project called "video".

Then create a new file and add the following definitions:

```
/*
**   NTSC Video
**   Graphic library
**
*/


#define VRES       192     // desired vertical resolution
#define HRES       256     // desired horizontal resolution (pixel)

void initVideo( void);

void haltVideo( void);

void clearScreen( void);

void synchV( void);

extern int VMap[HRES/16*VRES];
```

Save this file as "graphic.h" and add it to the same project.

Notice how the horizontal resolution and vertical resolution values are the only two parameters exposed. Within reasonable limits (due to timing constraints and the many considerations expressed in the previous sections), they can be changed to adapt to specific application needs; the state machine and all other mechanisms of the video generator module will adapt their timing as a consequence.

Testing the video generator

In order to test the video generator module we have just completed, we need only the MPLAB® SIM simulator tool and possibly a few lines of code for a main program:

```
//
// Graphic Test.c
//
// testing the basic graphic module
//

#include <p24fj128ga010.h>
#include "../graphic/graphic.h"

main()
{
    // initializations
    TRISA = 0xff80;        // set PORTA lsb as output for debugging
    clearScreen();         // init the video map
    initVideo();           // start the video state machine

    // main loop
    while( 1)
    {

    } // main loop

} // main
```

Save the project and use the build project checklist to build the entire project.

Open the logic analyzer window and use the logic analyzer checklist to add the RG0 pin (Sync) and the SDO1 output (Video) to the analyzer channels. At this point you could run the simulator for a few seconds and, after pressing the halt button, switch to the logic analyzer output window to observe the results. The trace memory of the simulator is of a limited capacity and can visuallize only small subset of an entire video frame. In other words, it is very likely that you will be confronted with a relatively uninteresting display containing a regular series of sync pulses and a flat video output. Unfortunately, the simulator does not simulate the output of the SPI port, so for that we'll have to wait until we run the application on real hardware. As per the Sync line, there is one interesting time we would like to observe—that is when we generate the vertical synchronization signal with a sequence of three long horizontal sync pulses at the beginning of each frame. By setting a breakpoint on the first line of the OC4 interrupt service routine (called for the first time at the beginning of the first image line), you can make sure that the simulation will stop relatively close to the beginning of a new frame.

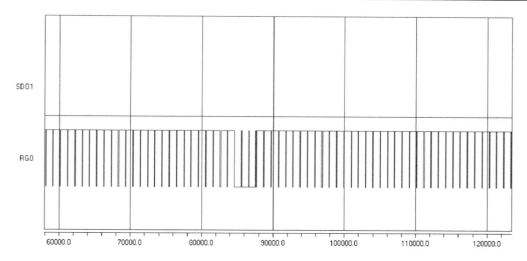

Figure 12-12. Screen capture of the logic analyzer window, vertical sync pulses.

If you are patient you can count the number of lines (one per sync pulse) following the three vertical sync (long) pulses and verify that they are in fact 33 (that is (262–192–3) / 2). Also you can zoom in the central portion to verify the proper timing of the sync pulses in the pre/post and vertical sync lines.

Using the cursors, you can verify the number of cycles composing a horizontal line period and the width of the horizontal sync pulse. Keep in mind that the logic analyzer window approximates the reading to the nearest screen pixel, so the accuracy of your reading will depend on the magnification (improving as you zoom in) and the resolution of your PC screen. Naturally if what you need is to determine a time interval with absolute precision, the most direct method is to use the stopwatch function of the MPLAB SIM software simulator together with the appropriate breakpoint settings.

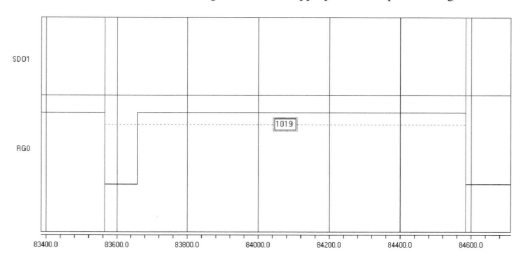

Figure 12-13. Zoomed view of a single pre-equalization line.

Measuring performance

Since the video generator module uses three different sources of interrupt and a state machine with four states, it might be interesting to get an idea of the actual processor overhead involved, possibly utilizing the logic analyzer to illustrate the percentage of time the processor spends inside the various interrupt service routines.

To this end, we will need to make a few simple modifications to all three of the interrupt service routines. We will use a pin of PORTA (RA0) as a flag that will be set to indicate when we are executing inside an interrupt service routine and cleared when we are executing the main loop:

```
void _ISRFAST _T3Interrupt( void)
{
    _RA0=1;
...
    _RA0=0;
} // T3Interrupt

void _ISRFAST _OC3Interrupt( void)
{
    _RA0=1;
...
    _RA0=0;
} // OC3Interrupt

void _ISRFAST _OC4Interrupt( void)
{
    _RA0=1;
...
    _RA0=0;
} // OC4Interrupt
```

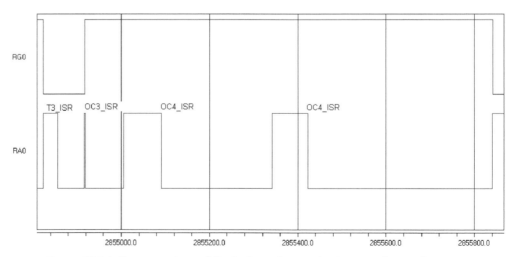

Figure 12-14. Screen capture of the logic analyzer output, measuring performance.

After recompiling and adding RA0 to the channels captured by the logic analyzer tool, we can zoom into a single horizontal line period (select an image line).

Using the cursors, we can measure the approximate duration of each interrupt service routine and, adding up the values for the worst possible case (an image line where all four interrupts are invoked), we obtain a value of 200 cycles out of a line period of 1018 cycles, representing an overhead of less than 20% of the processor time, a remarkably good result.

The dark screen

Playing with the simulator and the logic analyzer tool can be entertaining for a little while, but I am sure at this point you will feel an itch for the real thing! You can test the video interface on a real TV screen (or any other device capable of receiving an NTSC composite video signal) connected with the simple three-resistor interface to an actual PIC24 device. If you have an Explorer16 board, this is the time to take out the soldering iron and connect the three resistors to the small prototyping area in the top right corner of the demo board and out to a standard RCA video jack. Alternatively, if you feel your electronic hobbyist skills are up to the task, you could even develop a small PCB for a daughterboard that would fit in the expansion connectors of the Explorer16.

> Check on the companion web site "*www.flyingthepic24.com*" for the availability of expansion boards that will allow you to follow all the advanced projects of the third part of the book.

Whatever your choice, though, the experience will be breathtaking.

Or not! In fact, if you wire all the connections just right when you power up the Explorer16 board, what you are going to be staring at is just a blank (or I should better say black) screen. Sure, this is an achievement; in fact this already means that a lot of things are working right, as both the horizontal and vertical synchronization signals are being decoded correctly by the TV set and a nice, uniform black background is being displayed.

Figure 12-15. The dark screen.

A test pattern

To spice things up, we should start filling that video array with something worth looking at, possibly something simple that can give us an immediate feedback on the proper functioning of the video generator.

Let's create a new test program as follows:

```
//
// Graphic Test2.c
//
// testing the basic graphic module
//

#include <p24fj128ga010.h>
#include "../graphic/graphic.h"

main()
{
    int x, y;

    // fill the video memory map with a pattern
    for( y=0; y<VRES; y++)
        for (x=0; x<HRES/16; x++)
            VMap[y*16 + x]= y;

    initVideo();     // start the video state machine

    // main loop
    while( 1)
    {

    } // main loop

} // main
```

Instead of calling the clearScreen function, this time we use two nested `for` loops to initialize the VMap array. The external (*y*) loop counts the vertical lines, the internal (*x*) loop moves horizontally, filling the 16 words (each containing 16 bits) with the same value: the line count. In other words, on the first line each 16-bit word will be assigned the value 0, on the second line each word will be assigned the value 1, and so on until the last line (192) where each word will be assigned the value 191 (0xBF in hexadecimal).

If you build the new project and test the video output, you should be able to see the following pattern:

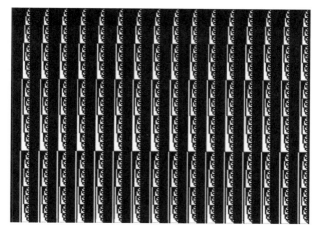

Figure 12-16. A screen capture of the video output generated with the test pattern.

In its simplicity, there is a lot we can learn from observing the test pattern. First of all, we notice that each word is visually represented on the screen in binary with the most significant bit presented on the left. This is a consequence of the order used by the SPI module to shift out bits: that is in fact msb first. Secondly, we can verify that the last row contains the expected pattern: 0x00bf, so we know that all rows of the memory map are being displayed. Finally, we can appreciate the detail of the image. Different output devices (TV sets, projectors, LCD panels,...) will be able to lock the image more or less effectively and/or will be able to present a sharper image depending on the actual display resolution and their input stage bandwidth. In general, you should be able to appreciate how the PIC24 can generate effectively straight vertical lines. This is not a trivial achievement. In fact, for each pixel to align properly row after row in a straight vertical line, there must be an absolute jitterless (deterministic) response to the timer interrupts, a notable characteristic of all PIC® microcontroller architectures.

This does not mean that on the largest screens you will not be able to notice small imperfections here and there, as small echoes and possibly minor visual artifacts in the output image. Realistically the simple three-resistor interface can only take us so far.

Ultimately, the entire composite video-signal interface could be blamed for a lower quality output. As you might know, S-Video, VGA and most other video interfaces keep luminance and synchronization signals separate to provide a more stable and clean picture.

Plotting

Now that we are reassured about the proper function of the graphic display module, we can start focusing more on generating the actual images onto the memory map. The first natural step is to develop a function that allows us to light up one pixel at a precise coordinate (x, y) on the screen. The first thing to do is derive the line number from the y coordinate. If the x and y coordinates are based on the traditional Cartesian plane representation, with the origin located in the bottom left corner of the screen, we need to invert the address before accessing the memory map, so that the first row in the memory map corresponds to the *y* maximum coordinate VRES-1 or 189, while the last row in the memory map corresponds to the *y* coordinate 0. Also, since our memory map is organized in rows of 16 words, we will need to multiply the resulting line number by 16 to obtain the address of the first word on the given line. This can be obtained with the following expression: VMap[(VRES-1 -y) *16].

Pixels are grouped in 16-bit words, so to resolve the *x* coordinate we first need to identify the word that will contain the desired pixel. A simple division by 16 will give us the word offset on the line. Adding the offset to the line address as calculated above will provide us with the complete word address inside the memory map:

```
VMap[ (VRES-1 -y) *16 + (x/16)]
```

In order to optimize the address calculation we can make use of shift operations to perform the multiplication and division as follows:

```
VMap[ (VRES-1 -y) << 4 + (x>>4)]
```

To identify the bit position inside the word corresponding to the required pixel, we can use the remainder of the division of *x* by 16, or more efficiently we can mask out the lower 4 bits of the *x* coordinate. Since we want to turn the pixel on, we will need to perform a binary OR operation with an appropriate mask that has a single bit set in the corresponding pixel position. Remembering that the display puts the msb of each word to the left (the SPI module shifts bits msb first) we can build the mask with the following expression:

```
( 0x8000 >> ( x & 0xf))
```

Putting it all together, we obtain the core plot function:

```
VMap[ ((VRES-1-y)<<4) + (x>>4)] |= (0x8000 >> (x & 0xf));
```

As a final touch we can add "clipping," that is a simple safety check, just to make sure that the coordinates we are given are in fact valid and within the current screen map limits:

```
void Plot( unsigned x, unsigned y)
{
    if ((x<HRES) && (y<VRES) )
        VMap[ ((VRES-1-y)<<4) + (x>>4)] |= (0x8000 >> (x & 0xf));
} // plot
```

By defining the *x* and *y* parameters as unsigned integers we guarantee that negative values will be discarded too, as they will be considered large integers outside the screen resolution.

A starry night

To test the newly developed plot function, we will create a new project. We will include the "graphic.c" and "graphic.h" files but we will also use the pseudo-random number generator functions available in the standard C library "stdlib.h". By using the pseudo-random number generator to produce random *x* and *y* coordinates for a thousand points, we will both test the plot function and, in a way, the random generator itself with the following simple code:

```
//
// Graphic Test3.c
//
// testing the basic graphic module
// plotting random points
//

#include <p24fj128ga010.h>

#include "../graphic/graphic.h"
#include <stdlib.h>


void plot( unsigned x, unsigned y)
{
    if ((x<HRES) && (y<VRES) )
        VMap[ ((VRES-1-y)<<4) + (x>>4)] |= (0x8000 >> (x & 0xf));
} // plot


main()
{
    int i;

    // initializations
    clearScreen();      // init the video map
    initVideo();        // start the video state machine

    srand(13);          // initialize the pseudo random number generator

    for( i=0; i<1000; i++)
    {
        plot( rand()%HRES, rand()%VRES);
    }

    // main loop
    while( 1)
    {

    } // main loop

} // main
```

The output on your video display should look like a nice starry night, as in the screen shot captured in Figure 12-17.

Figure 12-17. Screen capture, plotting a starry night.

A starry night it is, but not a realistic one you will notice, as there is no recognizable trace of any increased density of stars around a belt—in other words, there is no Milky Way! This is a good thing! This is a simple proof that our pseudo-random number generator is in fact doing the job it is supposed to.

We can now add the plot function to the "graphic.c" module. Remember to also add the prototype to the "graphic.h" function so that in the following exercises we will be able to use it.

```
void plot( unsigned, unsigned);
```

Line drawing

The next obvious step is drawing lines, or, I should better say, line segments. Granted, horizontal and vertical line segments are not a problem; a simple `for` loop can take care of it, but drawing oblique lines is a completely different thing. We could start with the basic formula for the line between two points that you will remember from the old school days:

```
y = y0 + (y1-y0)/(x1-x0) * ( x- x0)
```

where $(x0,y0)$ and $(x1,y1)$ are, respectively, the coordinates of two generic points that belong to the line.

This formula gives us, for any given value of x, a corresponding y coordinate, so we might be tempted to use it in a loop for each discrete value of x between the starting and ending point of the line, as in the following example:

```
//
// Line Test1.c
//
// testing the basic line drawing function
//

#include <p24fj128ga010.h>

#include "../graphic/graphic.h"
```

```
main()
{
    int x;
    float x0 = 10, y0 = 20, x1 = 200, y1 = 150, x2 = 20, y2 = 150;

    // initializations
    clearScreen();  // init the video map
    initVideo();    // start the video state machine

    // draw an oblique line (x0,y0) - (x1,y1)
    for( x=x0; x<x1; x++)
        plot( x, y0+(y1-y0)/(x1-x0)* (x-x0));

    // draw a second (steeper) line (x0,y0) - ( x2,y2)
    for( x=x0; x<x2; x++)
        plot( x, y0+(y2-y0)/(x2-x0)* (x-x0));


    // main loop
    while( 1)
    {

    } // main loop

} // main
```

The output produced is an acceptably continuous segment only for the first (shallower) line where the horizontal distance (x1 – x0) is greater than the vertical distance (y1 – y0). In the second, much steeper, line the dots appear disconnected and we are clearly unhappy with the result. Also we had to perform floating-point arithmetic, a computationally expensive proposition compared to integer arithmetic, as we have seen in previous chapters.

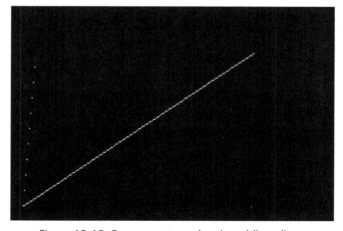

Figure 12-18. Screen capture, drawing oblique lines.

Bresenham algorithm

Back in 1962, when working at IBM in the San Jose development lab, Jack E. Bresenham developed a line-drawing algorithm that uses exclusively integer arithmetic and is today considered the foundation of any computer graphic program. Its approach is based on three optimization "tricks":

1. Reduction of the drawing direction to a single case (left to right).

2. Reduction of the line steepness to the single case where the horizontal distance is the greatest.

3. Multiplying both sides of the equation by the horizontal distance (`deltax`) to obtain only integer quantities.

The resulting line-drawing code is compact and extremely efficient; here is an adaptation for our video module:

```
#define abs( a)      (((a)> 0) ? (a) : -(a))

void line( int x0, int y0, int x1, int y1)
{
    int steep, t ;
    int deltax, deltay, error;
    int x, y;
    int ystep;

    steep = ( abs(y1 - y0) > abs(x1 - x0));

    if ( steep )
    { // swap x and y
        t = x0; x0 = y0; y0 = t;
        t = x1; x1 = y1; y1 = t;
    }
    if (x0 > x1)
    {  // swap ends
        t = x0; x0 = x1; x1 = t;
        t = y0; y0 = y1; y1 = t;
    }

    deltax = x1 - x0;
    deltay = abs(y1 - y0);
    error = 0;
    y = y0;

    if (y0 < y1) ystep = 1; else ystep = -1;
    for (x = x0; x < x1; x++)
    {
        if ( steep) plot(y,x); else plot(x,y);
        error += deltay;
        if ( (error<<1) >= deltax)
        {
```

```
                    y += ystep;
                    error -= deltax;
             } // if
        } // for
} // line
```

We can add this function to the video module "graphic.c" and a prototype to the include file "graphic.h".

To test the efficiency of the Bresenham algorithm, we can create a new small project and once more include the pseudo-random number generator by including the "stdlib.h" library. The following example code will first draw a frame around the screen and then it will exercise the line-drawing routine producing a hundred lines from randomly generated coordinates. The main loop also contains a check for the S3 button (the leftmost button on the bottom of the Explorer16 demo board) to be pressed before the screen is cleared again and a new set of random lines is drawn on the screen.

```
//
// Bresenham.c
//
// Bresenham algorithm example
//

#include <p24fj128ga010.h>
#include <stdlib.h>
#include "../graphic/graphic.h"

main()
{
    int i;

    // initializations
    initVideo();     // start the state machines
    srand( 12);

    // main loop
    while( 1)
    {
        clearScreen();
        line( 0, 0, 0, VRES-1);
        line( 0, VRES-1, HRES-1, VRES-1);
        line( HRES-1, VRES-1, HRES-1, 0);
        line( 0, 0, HRES-1, 0);

        for( i = 0; i<100; i++)
            line( rand()%HRES, rand()%VRES, rand()%HRES, rand()%VRES);

    // waiting for a button to be pressed
        while( 1)
```

```
        {
            if ( !_RD6)
                break;
        } // wait

    } // main loop

} // main
```

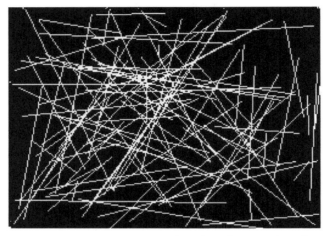

Figure 12-19. Screen capture, Bresenham line-drawing algorithm test.

You will be impressed by the speed of the line-drawing algorithm; even when increasing the number of lines drawn to batches of one thousand, the performance of the PIC24 will be apparent.

Plotting math functions

With the completed graphic module we can now start exploring some interesting applications that can take full advantage of its visualization capabilities. One classical application could be plotting a graph based on data logged from a sensor or, more simply for our demonstration purposes, calculated on the fly from a given math function.

For example, let's assume the function is a sinusoid with a twist as in the following:

$$y\,(x) = x * \sin(\,x\,)$$

Let's also assume we want to plot its graph for values of x between 0 and 8 * PI.

With minor manipulations we can scale the function to fit our screen, remapping the input range from 0 to 200 and the output range to the +75/–75 values range.

The following program example will plot the function after tracing the *x* and *y* axes:

```c
/*
** Plotting a 1D function graph
**
*/

#include <p24fj128ga010.h>
#include <math.h>

#include "../graphic/graphic.h"

#define X0 10
#define Y0 (VRES/2)
#define PI 3.141592654f

main( void)
{
    int x, y;
    float xf, yf;

    // initializations
    clearScreen();
    initVideo();

    // draw the x and y axes crossing in (X0,Y0)
    line( X0, 10, X0, VRES-10);     // y axes
    line( X0-5, Y0, HRES-10, Y0);   // x axes

    // plot the graph of the function for
    for( x=0; x<200; x++)
    {
        xf = (8 * PI / 200) * (float) x;
        yf =  75.0 / ( 8 * PI) * xf * sin( xf);
        plot( x+X0, yf+Y0);
    }

    // main loop
    while( 1);

} // main
```

Should the points on the graph become too sparse, we have the option of using the line-drawing algorithm to connect each point to the previous.

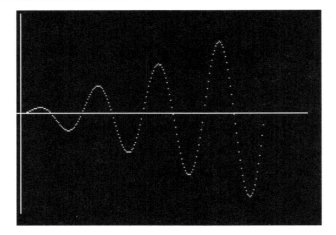

Figure 12-20. Screen capture, a sinusoidal function graph.

Two-dimensional function visualization

Plotting two-dimensional function graphs could be more interesting and perhaps entertaining. This adds the thrill of managing the perspective distortion and the challenge of connecting the calculated points to form a visually pleasing grid.

The simplest method to squeeze the third axis into a two-dimensional image is to utilize what is commonly known as an isometric projection, a method that requires minimal computational resources while providing a small visual distortion. The following formulas applied to the x, y and z coordinates of a point in a three-dimensional space produce the px and py coordinates of the projection on a two-dimensional space (our video screen).

$px = x + y/2;$

$py = z + y/2;$

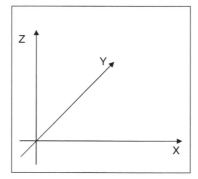

Figure 12-21. Isometric projection.

In order to plot the three-dimensional graph of a given function: $z = f(x,y)$ we proceed on a grid of points equally spaced in the x and y plane using two nested `for` loops. For each point we compute the function to obtain the *z* coordinate and we apply the isometric projection to obtain a (*px,py*) coordinate pair. Then we connect the newly calculated point with a segment to the previous point on the same row (previous column). A second segment needs to be drawn to connect the point to the previously computed point in the same column and the previous row.

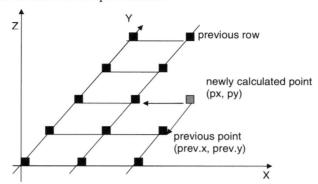

Figure 12-22. Drawing a grid to enhance a two-dimensional graph visualization.

While it is trivial to keep track of the coordinates of the previously computed point on the same row, recording the coordinates of the points on each previous row might require significant memory space. If, for example, we are using a grid of 20 × 20 points, we would need to store the coordinates of up to 400 points. Requiring two integers each, that would add up to 800 words or 1,600 bytes of precious RAM memory. In reality, as should be evident from the picture above, all we really need is the coordinates of the points on the "edge" of the grid as painted so far. Therefore, with a little care, we can reduce the memory requirement to just 20 coordinate pairs by maintaining a small rolling buffer.

The example code below visualizes the graph of the function:

$$z(x,y) = 1/ \text{sqrt}(x^2 + y^2) * \cos (\text{sqrt}(x^2 + y^2)$$

for values of *x* and *y* in the range –3 * PI to +3 * PI

```
/*
** Plotting a 2D function graph
**
**
*/

#include <p24fj128ga010.h>
#include <math.h>

#include "../graphic/graphic.h"

#define X0      10
#define Y0      10
#define PI      3.141592654f
#define NODES   20
#define SIDE    10
```

```
typedef struct {
        int x;
        int y;
    } point;

point edge[NODES], prev;

main( void)
{
    int i, j, x, y, z;
    float xf, yf, zf, sf;
    int px, py;

    // initializations
    clearScreen();
    initVideo();

    // draw the x, y and z axes crossing in (X0,Y0)
    line( X0, 10, X0, VRES-50);          // z axis
    line( X0-5, Y0, HRES-10, Y0);        // x axis
    line( X0-2, Y0-2, X0+120, Y0+120);   // y axis

    // init the array of previous egde points
    for( j = 0; j<NODES; j++)
    {
        edge[j].x = X0+ j*SIDE/2;
        edge[j].y = Y0+ j*SIDE/2;
    }

    // plot the graph of the function for
    for( i=0; i<NODES; i++)
    {
        // transform the x coordinate range to 0..200 offset 100
        x = i * SIDE;
        xf = (6 * PI / 200) * (float)(x-100);
        prev.y = Y0;
        prev.x = X0 + x;

        for ( j=0; j<NODES; j++)
        {
            // transform the y coordinate range to 0..200 offset 100
            y = j * SIDE;
            yf = (6 * PI / 200) * (float)(y-100);

            // compute the function
            sf = sqrt( xf * xf + yf * yf);
            zf = 1/(1+ sf) * cos( sf );
```

234

```
        // scale the output
        z = zf * 75;

        // apply isometric perspective and offset
        px = X0 + x + y/2;
        py = Y0 + z + y/2;

        // plot the point
        plot( px, py);

        // draw connecting lines to visualize the grid
        line( px, py, prev.x, prev.y);  // connect to prev point on same x
        line( px, py, edge[j].x, edge[j].y);

        // update the previous points
        prev.x = px;
        prev.y = py;
        edge[j].x = px;
        edge[j].y = py;
    } // for j
} // for i

// main loop
while( 1);

} // main
```

After building the project and connecting to a display, you will notice how quickly the PIC24 will produce the output graph, although significant floating-point math is required as the function is applied sequentially to 400 points and as much as 800 line segments are drawn on the video memory.

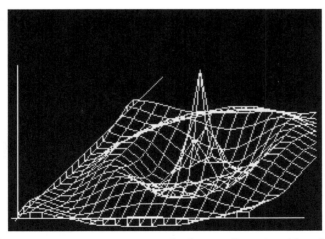

Figure 12-23. Screen capture, graph of a two-dimensional function.

Fractals

"Fractals" is a term coined for the first time by Benoit Mandelbrot, a mathematician (and fellow researcher at the IBM Pacific Northwest Labs) back in 1975, to denote a large set of mathematical objects which presented an interesting property: that of appearing self-similar at all scales of magnification as if constructed recursively with an infinite level of detail. There are many examples of fractals in nature, although their self-similarity property is typically extended over a finite scale range. Examples include clouds, snow flakes, mountains, river networks and even the blood vessels in our body.

Because it lends itself to impressive computer visualizations, the most popular example of a mathematical fractal object is perhaps the Mandelbrot set. It's defined as a subset of the complex plane where the quadratic function $z^2 + c$ is iterated. By exclusion, points (c) of the complex plane for which the iteration does not "diverge" are considered to be part of the set. Since it is easy to prove that once the modulus of z is greater than 2, the iteration is bound to diverge (hence the given point is not part of the set) we can proceed by elimination. The problem is that as long as the modulus of z remains smaller than 2, we have no way of telling when to stop the iteration and declare the point part of the set. So typically computer algorithms that depict the Mandelbrot set use an approximation, by setting an arbitrary maximum number of iterations past which a point is simply assumed to be part of the set.

Here is an example of how the inner iteration can be coded in C:

```
// initialization
  x = x0;
  y = y0;
  k = 0;

 // core iteration
 do {
    x2 = x*x;
    y2 = y*y;
    y = 2*x*y + y0;
    x = x2 - y2 + x0;
    k++;
  } while ( (x2 + y2 < 4)  && ( k < MAXIT));

// check if the point belongs to the Mandelbrot set
if ( k == MAXIT) plot( x0, y0);
```

where x0 and y0 are the coordinates in the complex space of the point c.

We can repeat this iteration for each point of a squared subset of the complex plane so as to obtain an image of the entire Mandelbrot set. From the literature we learn that the entire set is included in a disc of radius 2 around the origin, so we can develop a first program that will scan the complex plane in a grid of 192 × 192 points (to use the maximum screen resolution as defined by our video module) overlapping such a disc:

```
/*
**
** Mandelbrot Set graphic demo
**
*/
```

```c
#include <p24fj128ga010.h>
#include "../graphic/graphic.h"

#define SIZE   VRES
#define MAXIT 64

void mandelbrot( float xx0, float yy0, float w)
{
    float x, y, d, x0, y0, x2, y2;
    int i, j, k;

    // calculate increments
    d = w/SIZE;

    // repeat on each screen pixel
    y0 = yy0;
    for (i=0; i<SIZE; i++)
    {
        x0 = xx0;
        for (j=0; j<SIZE; j++)
        {
            // initialization
             x = x0;
             y = y0;
             k = 0;

             // core iteration
             do {
               x2 = x*x;
               y2 = y*y;
               y = 2*x*y + y0;
               x = x2 - y2 + x0;
               k++;
             } while ( (x2 + y2 < 4)  && ( k < MAXIT));

            // check if the point belongs to the Mandelbrot set
            if ( k == MAXIT) plot( j, i);

            // compute next point x0
            x0 += d;
        } // for j
        // compute next y0
        y0 += d;
    } // for i
  } // mandelbrot
```

```
 main()
{
    float x, y, w;

    // initializations
    initVideo();        // start the state machines

    // intial coordinates lower left corner of the grid
    x = -2.0;
    y = -2.0;
    // initial grid side
    w =   4.0;

    while( 1)
    {

        clearScreen();   // clear the screen
        mandelbrot( x, y, w);

        while (1);

    } // main loop

} // main
```

With the maximum number of iterations set to 64, the PIC24 will produce the complete image below, the so-called Mandelbrot cardioid, in approximately 30 seconds.

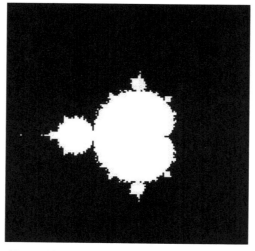

Figure 12-24. Screen capture, Mandelbrot set.

I will confess that since I bought my first personal computer as a kid (actually "home computer" was the term used back then—it was a Sinclair ZX Spectrum), I have been playing with fractal programs. So I have a vivid memory of the long hours I used to spend staring at the computer screen waiting for the old trusty Z80 processor (running at the whopping speed of 3.5 MHz) to paint this same image. A few years later, my first IBM PC, an XT clone (running on a 8088 processor at a not much higher clock speed of 4 MHz) was not faring much better, and although the screen resolution of my monochrome Hercules graphic card was higher, I would still launch programs in the evening to watch the results the following morning, after what amounted sometimes to up to eight hours of processing. Clearly the amount of computation required to paint a fractal image varies enormously with the chosen area and the number of maximum iterations allowed, but the first time I ran this program I could not help being amazed by how rapidly the PIC24 painted the cardioid before my eyes.

But the real fun has just begun. The most interesting parts of the Mandelbrot set are at the fringes, where we can increase the magnification and zoom in to discover an infinitely complex world of details. By visualizing not just the points that belong to the set, but also the ones at its edges that diverge, and assigning a color that depends on how fast they did in fact diverge, we can further improve the resulting image. Since we have only a monochrome display, we will simply use alternate bands of black and white assigned to each point according to the number of iterations it took before it either reached the maximum modulus or the maximum number of iterations. Simply enough, this means we will have to modify just one line of code from our previous example:

```
    . . .
            // check if the point belongs to the Mandelbrot set
            if ( k & 1) plot( j, i);
    . . .
```

Also, since the best way to play with Mandelbrot set images is to explore them by selecting new areas and zooming in on the details, we can transform the main program loop by adding a simple user interface, by means of the four buttons of the Explorer16 board. We can imagine splitting the image into four quadrants. A button will correspond to each quadrant, and by pressing it, we will zoom in, doubling the resolution and halving the grid dimension (w).

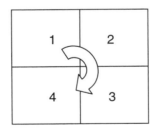

Figure 12-25. Splitting the screen into four quadrants.

```
main()
{
    float x, y, w;

    // initializations
    initVideo();      // start the state machines

    // intial coordinates lower left corner of the grid
    x = -2.0;
    y = -2.0;
    // initial grid size
    w =  4.0;

    while( 1)
    {
        clearScreen();                           // clear the screen
        mandelbrot( x, y, w);        // draw new image

        // wait for a button to be pressed
        while (1)
        {  // wait for a key pressed
            if ( !_RD6)
            {  // first quadrant
                w/= 2;
                y += w;
                break;
            }
            if ( !_RD7)
            { // second quadrant
                w/= 2;
                y += w;
                x += w;
                break;
            }
            if ( !_RA7)
            {  // third quadrant
                w/= 2;
                x += w;
                break;
            }
            if ( !_RD13)
            { // fourth quadrant
                w/= 2;
                break;
            }
        } // wait for a key
    } // main loop
} // main
```

Here is a little selection of interesting areas you will be able to explore with a little patience:

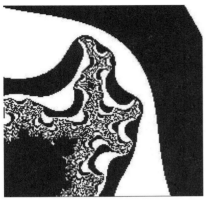

Figure. 12-26a
(+0.25 +*j* 0.5), *w* = 0.25

Figure. 12-26b
(+0.37500 −*j* 0.57813), *w* = 0.01563

Figure 12-26c
(−1.28125 +*j* 0.3125), *w* = 0.3125

Figure. 12-26d
(+0.34375 +*j* 0.56250), *w* = 0.03125

Figure. 12-26e
(+0.34375 +*j* 0.56250), *w* = 0.03125

Text

So far we have been focusing heavily on graphical visualizations, but on more than one occasion you might have felt the desire to augment the information presented on the screen with some text. Writing text on the video memory is no different from plotting points or drawing lines, and in fact it can be achieved with a variety of methods, including using the plotting and line-drawing functions we have already developed. But for greater performance and in order to require the smallest possible amount of code, the easiest way to get text on our graphic display is by developing and using an 8 × 8 font array. Each character can be drawn in an 8 × 8 pixel box; one byte will encode each row, and 8 bytes will encode the entire character. We can then assemble the 96 base alphabetical, numerical and punctuation characters in the order and position in which they are presented in the ASCII character set in a single array and save it as an include file.

0	0	0	1	1	1	0	0
0	0	1	0	0	0	1	0
0	0	1	0	0	0	1	0
0	0	1	1	1	1	1	0
0	0	1	0	0	0	1	0
0	0	1	0	0	0	1	0
0	0	1	0	0	0	1	0
0	0	0	0	0	0	0	0

Figure 12-27. The letter A as represented in an 8 × 8 font.

To save space we don't need to create the first 32 codes defined in the ASCII set, which correspond mostly to commands and legacy special synchronization codes used by teletypewriters and modems of the old times:

```
//
// 8 x 8 Font definition
//

#define F_OFFS  0x20     // initial offset
#define F_SIZE  0x60     // only the first 64 characters defined so far

const char Font8x8[] = {
// 20 - SPACE
    0b00000000,
    0b00000000,
    0b00000000,
    0b00000000,
    0b00000000,
    0b00000000,
    0b00000000,
```

```
    0b00000000,
// 1 - !
    0b00011000,
    0b00011000,
    0b00011000,
    0b00011000,
    0b00011000,
    0b00000000,
    0b00011000,
    0b00000000,
...
```

Notice that the `Font8x8[]` array is defined with the attribute `const`, as its contents are supposed to remain unchanged during the execution of the program and it is best allocated in the program memory space (Flash memory of the PIC24) to save precious RAM memory space.

A complete listing of the "`font.h`" file would waste several pages, so we will omit it here, but you will be able to find it on the companion CD-ROM.

Of course, as it is a matter of personal taste, you are welcome to modify the `Font8x8[]` array contents to fit your preferences.

Printing a character on the screen then is a matter of copying one byte at a time from the font array to the desired screen position. In the simplest case, characters can be aligned to the words that compose the VMap (video memory) array defined by the graphics module. In this way, the character positions would be limited to 32 characters per line (256/8) and a maximum of 24 rows of text could be displayed (192/8). A more advanced solution would call for absolute freedom in positioning each character at any given pixel coordinate. This would require a type of manipulation often referred to as BitBLT (an acronym that stands for Bit BLock Transfer) that is common in computer graphics, particularly in video game design. In the following, though, we will stick to the simpler approach, looking for the solution that requires the smallest amount of resources to get the job done.

Let's create a new project that we will call "TextOnGPage" and a new source file "`TextOnGPage.c`" that will contain all the functions required to print text on the graphic video page. Then, let's define two integer variables for maintaining the cursor position:

```
int cx, cy;
```

We can now write a simple function that prints one ASCII character at a time on the screen at the current cursor position as follows:

```
void putcV( int a)
{
    int i, *p;
    const char *pf;

    // 1. check if char in range
    a -= F_OFFS;
    if ( a < 0)        a = 0;
    if ( a >= F_SIZE)  a = F_SIZE-1;
```

```
    // 2. check page boundaries
    if ( cx >= HRES/8)              // wrap around x
    {
        cx = 0;
        cy++;
    }
    if ( cy >= VRES/8)              // wrap around y
        cy = 0;

    // 3. set pointer to word in the video map
    p = &VMap[ cy * 8 * HRES/16 + cx/2];
    // set pointer to first row of the character in the font array
    pf = &Font8x8[ a << 3];

    // 4. copy one by one each line of the character on the screen
    for ( i=0; i<8; i++)
    {
        if ( cx & 1)
        {
            *p &= 0xff00;
            *p |= *pf++;
        }
        else
        {
            *p &= 0xff;
            *p |= (*pf++)<<8;
        }
        // point to next row
        p += HRES/16;
    } // for

    // increment cursor position
    cx++;
} // putcV
```

In the very first few lines of the function (1.) we verify that the character passed to the function is part of the subset of the ASCII character set currently defined in our font. If not, we change it into either the first character defined or the last one. An alternative strategy, available to the reader, would have been to ignore the character altogether and exit the routine immediately in such a case.

The second part of the function (2.) deals with positioning the cursor (cy, cy), making sure that if we reach the right edge of the screen we wrap around onto the next line as a typewriter would. A similar action is taken when we reach the bottom right extreme of the screen by wrapping around to the top of the screen. The alternative here would have been to implement a scrolling feature that would move the entire contents of the screen up by one line to make room for a whole new line of text.

In the third part (3.) a pointer to the screen memory map is computed based on the cursor coordinates, and a pointer into the font array is computed based on the ASCII character code. Finally (4.) a loop takes care of copying, line by line, the font image into the video array. Since the video array (VMap) is organized in words (and the MSB is displayed first) a little attention must be paid in transferring each byte to the proper position within each 16-bit word. If the cursor position is even, the MSB of the selected word is replaced by the font data. If the cursor position is odd, the LSB of the selected word is replaced by the font data. At each step in the loop, the pointer inside the video map (p) is incremented by 16 words (HRES/16) to point to the same position on the following line, while the pointer inside the font array (pf) is incremented by one to obtain the next byte composing the character image.

For our convenience we can now create a function that will print an entire NULL terminated ASCII string on the screen:

```
void putsV( unsigned char *s)
{
    while (*s)
        putcV( *s++);
} // putsV
```

Remember also to include all the necessary files to compile this module:

```
#include <p24fj128ga010.h>
#include "../font/font.h"
#include "../graphic/graphic.h"
```

Finally, let's create a new include file to export the newly defined functions and to add a couple of useful macros:

```
/*
** Text on Graphic Page
*/

extern int cx, cy;

void putcV( int a);

void putsV( unsigned char *s);

#define Home()      { cx=0; cy=0;}
#define Clrscr()    { clearScreen(); Home();}
#define AT( x, y)   { cx = (x); cy = (y);}
```

Home() will simply position the cursor on the upper left corner of the screen.

Clrscr() will clear the screen first by invoking the function defined in the graphic module.

AT() will position the cursor as required for the next putcV and/or putsV command.

Notice how, differently from the graphic coordinate system, the text cursor coordinate system is defined with the origin located in the home position on the upper left corner of the screen and increasing vertical coordinates are referring to lines further down the page.

Testing the TextOnGPage module

In order to quickly test the effectiveness of the new text module, we can now create a small program that, after printing a small banner on the first line of the screen will print out each character defined in the 8×8 font:

```
/*
** Text Page Test
**
*/

#include <p24fj128ga010.h>
#include "../graphic/graphic.h"
#include "../textg/TextOnGPage.h"


main( void)
{
    int i;

    // initializations
    initVideo();    // start the state machines

    Clrscr();

    AT( 0, 0);
    putsV( "FLYING THE PIC24!");

    AT( 0, 2);
    for( i=32; i<128; i++)
    putcV( i);

    while (1);

} // main
```

Save this file as "TextOnGTest.c" and add it to the project. Make sure all the other required modules are added to the project too, including: "graphic.c", "graphic.h", "font.h", "textongpage.c" and "textongpage.h". Finally, build the project and run.

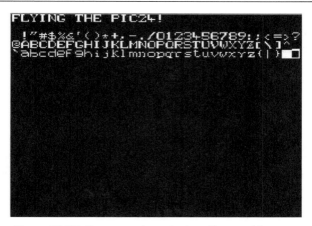

Figure 12-28. Screen capture, text on the graphic page.

Developing a text page video

Using the newly developed "TextOnGPage.c" module, we have now acquired the capability to display text and graphics on the video screen. The system in its entirety requires 6,080 bytes of RAM for the video map, a significant portion of the total amount of RAM available inside the PIC24fj128ga010, but only a minuscule portion of the program memory available.

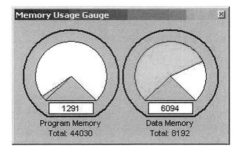

Figure 12-29. Memory usage gauges for the TextOnGTest project.

If our application was going to need the video output only to display text, this would have been an extremely inefficient solution. In fact, using an 8×8 font, we can only display 32 characters per line and a maximum of 24 lines, for a grand total of 768 characters. In other words, if our application uses the video as a pure text display, we are wasting as much as 5,244 bytes of precious RAM. In the early computer days (including the first IBM PC) this was a serious (economical) problem that demanded a custom hardware solution. All early personal computer systems had in fact a "text page," that is a video mode where the display could visualize ONLY text, with the advantage of reducing considerably the RAM requirements (to a fraction of those of a graphic page) while also increasing considerably the screen manipulation performance. In a text page, character ASCII codes are stored directly in the video memory and they are converted on the fly to the graphical font representation by a hardware device (known as the font generator) intimately connected to the video scanning and timing logic. In this way, the amount of memory required to maintain a page of 768 characters (as in our previous project) would have been only and exactly 768 bytes; that is approximately only 10% of the memory required by our graphic-display solution.

This sounds to me like an interesting new challenge. In the next project we will develop a more RAM-efficient video solution targeting pure text display applications. This will force us to go back to the initial definition of the state machine at the heart of the graphic video module. In fact, we can keep most of its structure intact and proceed to optimize only a few critical areas. All the elements that compose the horizontal and vertical synchronization signals will remain unchanged. Also the construction of horizontal lines remains untouched up to the point where we start sending data to the SPI1 module to serialize. Where in the graphic display we take each word of the memory map as is and we push it on to the SPI buffer, in a text-page video application we will need to operate on a byte at a time and interpose a conversion step. The Font8x8[] array will act as a look-up table that will be used to convert the ASCII code on the fly from the text page (now VMap will be defined as a byte array) into an image that will be sent to the SPI buffer for serialization. In generic terms we can express this translation with the following expression:

```
lookup = Font8x8[ *VPtr * 8  + RCount];
```

where VPtr is the pointer to the current character inside the text page array, and RCount is a counter from 0 to 7 that keeps track of each video line forming one row of text (there are 8 video lines for each row of text).

In practice, things are a little more complicated. Since the SPI module must be fed with 16 bits of data at a time, we need to assemble two characters in one word after performing two look-ups one after the other:

```
lookup1 = Font8x8[ *VPtr++ * 8  + RCount];
lookup2 = Font8x8[ *VPtr++ * 8  + RCount];
SPI1BUF = ( 256 * lookup1 + lookup2);
```

Repeat this for 8 times to fill the entire SPI buffer.

Now this is a lot of work to perform in the few microseconds available to the OC4 interrupt service routine. Even if we were to enable the highest level of optimization of the compiler (and in this book we actually chose to never enable any optimization), the possibility that it would fit in the time available (less than 25 µs) is pretty slim. There are simply too many multiplications and additions to perform when working the look-up table. Fortunately, this is something we can change. In fact we can rearrange the way the Font array is built. While it is convenient to initialize the array filling in all 8 rows of each character and proceeding sequentially, in order to simplify the look-up expression it would be best if the array were organized the other way around. In other words, we should fill the array starting with the first byte of each character in the font, followed by each second byte of each character and so on. We could rewrite the expressions above with the new rearranged font RFont as follows:

```
lookup1 = RFont[  (RCount * F_SIZE) + *VPtr++ ];
lookup2 = RFont[  (RCount * F_SIZE) + *VPtr++ ];
SPI1BUF = ( 256 * lookup1 + lookup2);
```

The great advantage lies in the fact now (RCount*F_SIZE) is a constant offset and we can even obtain a pointer inside the font that already takes care of such offset with the following expression:

```
FPtr = &RFont[ RCount * F_SIZE];
```

This can be precalculated (inside the Timer3 interrupt service routine) at the beginning of each line for a significant saving.

The new look-up expressions are now simplified to:

```
lookup1 = FPtr[ *VPtr++ ];
lookup2 = FPtr[ *VPtr++ ];
SPI1BUF = ( lookup1 << 8 + lookup2);
```

Now we at least have a chance that the look-up expression could fit in the few microseconds available, but we are not satisfied yet. Every nanosecond counts in a routine as critical and frequently invoked as the OC4 interrupt service routine is. The ultimate optimization trick is in fact constituted by the selective manual coding in assembly of the few most critical steps. If we assume that the font pointer (FPtr) has been placed in the W2 working register and the video memory pointer (VPtr) has been placed in the W1 working register, we can code the entire look-up sequence with just three powerful assembly instructions:

```
mov.b [w1++], w0      // w0 = *VPtr++     (8 bit)
ze w0, w0             // extend w0 to a 16 bit integer
mov.b [w2+w0], w3     // w3 = FPtr[ w0] = FPtr[ *VPtr++] = lookup1
```

Repeating the same instructions for lookup2 is trivial; combining the two values in one word requires only a shift:

```
sl w3, #8, w3         // shift W3 8 bits to the left (*256)
```

and later on an addition:

```
add w0, w3, w0        // add (lookup1*256) and lookup2
```

We can put it all together in a single macro that we will call DECODE():

```
#define DECODE( sfr) \
    asm volatile ( "mov.b [w1++], w0" );      \
    asm volatile ( "ze w0, w0");              \
    asm volatile ( "mov.b [w2+w0], w3");      \
    asm volatile ( "sl w3,#8,w3" );           \
    asm volatile ( "mov.b [w1++], w0" );      \
    asm volatile ( "ze w0, w0");              \
    asm volatile ( "mov.b [w2+w0], w0");      \
    asm volatile ( "ze w0, w0");              \
    asm volatile ( "add w0, w3, w0");         \
    asm volatile ( "mov w0, %0" : "=U"((sfr)));
```

The volatile attribute is used here to make sure that the compiler will not change the order and position of the inline assembly code should the optimizer be turned on in the future. Also, the last line might seem a bit cryptic. In fact, we are using an advanced feature of the inline assembly syntax offered by the C30 compiler that allows us to mix in C variable names, passed as parameters to the asm() function. The special notation :"=U"() indicates that a data operand in brackets is being passed as an output data recipient.

We can now modify the OC4 interrupt routine to make full use of our highly optimized font table lookup:

```
void _ISRFAST _OC4Interrupt( void)
{
    // prepare pointers
    volatile asm ( "mov %0, w2" ::"U" (FPtr) );    // w2 = FPtr
    volatile asm ( "mov %0, w1" ::"U" (VPtr));     // w1 = VPtr

    // inline text to font translation * 8 words
    DECODE( SPI1BUF);
    DECODE( SPI1BUF);
    DECODE( SPI1BUF);
    DECODE( SPI1BUF);
    DECODE( SPI1BUF);
    DECODE( SPI1BUF);
    DECODE( SPI1BUF);
    DECODE( SPI1BUF);
    __asm__( "mov w1, %0" :"=U" (VPtr));    // update VPtr

    if ( --HCount > 0)
    {   // activate again in time for the next SPI load
            OC4R += ( PIX_T * 8 * 16);
            OC4CON = 0x0009;    // single event
    }

    // clear the interrupt flag
    _OC4IF = 0;

} // OC4Interrupt
```

As we said before, the modifications to the Timer3 interrupt service routine are minor, as only a couple of pointers need to be prepared for the text lines to be properly sequenced and for the font offset to be precalculated:

```
void _ISRFAST _T3Interrupt( void)
{
    // Start a Sync pulse
    SYNC = 0;

    // decrement the vertical counter
    VCount--;

    // vertical state machine
    switch ( VState) {
        case SV_PREEQ:
            // horizontal sync pulse
            OC3R = HSYNC_T;
            OC3CON = 0x0009;    // single event
            break;
```

```
        case SV_SYNC:
            // vertical sync pulse
            OC3R = H_NTSC - HSYNC_T;
            OC3CON = 0x0009;    // single event
            break;

        case SV_POSTEQ:
            // horizontal sync pulse
            OC3R = HSYNC_T;
            OC3CON = 0x0009;    // single event
            // on the last posteq prepare for the new frame
            if ( VCount == 0)
            {
                LPtr = VMap;
                RCount = 0;
            }
            break;

        default:
        case SV_LINE:
            // horizontal sync pulse
            OC3R = HSYNC_T;
            OC3CON = 0x0009;    // single event

            // activate OC4 for the SPI loading
            OC4R = HSYNC_T + BPORCH_T;
            OC4CON = 0x0009;    // single event
            HCount = 3;         // reload counter

            // prepare the font pointer
            FPtr = &RFont[ RCount * F_SIZE];
            // prepare the line pointer
            VPtr = LPtr;

            // Advance the RCount
            if ( ++RCount == 8)
            {
                RCount = 0;
                LPtr += COLS;
            }
} //switch

// advance the state machine
if ( VCount == 0)
{
    VCount = VC[ VState];
```

```
        VState = VS[ VState];
    }


    // clear the interrupt flag
    _T3IF = 0;


} // T3Interrupt
```

The video initialization routine will now require one more step, as the font array needs to be rearranged as discussed above:

```
// prepare a reversed font table
for (i=0; i<8; i++)
{
    p = Font8x8 + i;
    for (j=0; j<F_SIZE; j++)
    {
        *r++ = *p;
        p+=8;
    } // for j
}   // for i
```

While for simplicity we implement this as a second array allocated in RAM where we copy things in the new order, the ultimate solution is to rearrange the "font.h" file definition, so that the Font8x8 array is already defined in the new and optimal order, there is no RAM waste, and no processing time is used during the video initialization to perform the translation.

Back when we were working on the graphical interface, we found that a 256 × 192 pixel screen was an acceptable compromise between screen resolution and memory usage as it would leave 2 kbytes of RAM available for the application to use. Now the balance is considerably changed; with a 24 lines by 32 column display, only 768 bytes are used by the video module and we can in fact afford to expand the resolution a bit. The horizontal resolution is the one most in need of an upgrade. Most video terminals use a 25 × 80 format while the average printed document has no less than 60 characters per line. While we could afford the RAM (25 rows × 80 columns = 2,000 characters), this time it is the NTSC video specifications that are going to dictate the ultimate limit. As we observed at the very beginning of this chapter, the maximum signal bandwidth for an NTSC video composite signal is fixed at 4.2 MHz, while the portion of the waveform producing the visible line image is 52 µs wide. This determines a maximum theoretical horizontal resolution of 436 pixels that, in the case of an 8×8 font, would imply a maximum number of 54 columns. In practice we would do better to choose a smaller value and, to make the best use of the SPI FIFO mechanism that we have been using with success so far, we had better choose a number that is a multiple of 16. While in the graphic module we used two successive blocks of 128 pixels each to fill the SPI FIFO buffers, for the text page module we can now add a third block, bringing the total horizontal resolution up to 48 characters. Note how this will require the SPI clock prescaler to be switched to the higher frequency mode (PIX_T= 2).

For the vertical resolution we have considerable freedom, since the NTSC standard specifies 262 lines of which theoretically up to 253 could be used for the actual image. There is no difficulty in making 25 rows of text (adding up to 200 lines) fit.

Overall our text-page module will produce a 25 row by 48 column display, using a total of just 1,200 bytes. This will represent a considerable improvement in readability with respect to the text on graphic page approach, with a significant reduction in the RAM memory usage as well.

This is the new set of constants and definitions that completes the new "Text" video module:

```c
/*
** TextPage.c
**
** Text Page video module
**
*/

#include <p24fj128ga010.h>
#include "../Text/TextPage.h"
#include "../font/font.h"

// I/O definitions
#define SYNC     _LATG0  // output
#define SDO      _RF8    // SPI1 SDO

// calculates the NTSC video parameters for the vertical state machine
#define V_NTSC    262    // total number of lines composing a frame
#define VRES     (ROWS*8) // desired vertical resolution (<242)
#define VSYNC_N  3       // V sync lines
// count the number of remaining black lines top+bottom
#define VBLANK_N    (V_NTSC -VRES - VSYNC_N)

#define PREEQ_N    VBLANK_N /2         // pre equalization + bottom blank
#define POSTEQ_N   VBLANK_N - PREEQ_N  // post equalization + top blank lines

// definition of the vertical sync state machine
#define SV_PREEQ    0
#define SV_SYNC     1
#define SV_POSTEQ   2
#define SV_LINE     3

// calculates the NTSC video parameters for the horizontal state machine
#define H_NTSC   1018    // total number of Tcy in a line (63.5us)
#define HRES    (COLS*8) // desired horizontal resolution (divisible by 16)
#define HSYNC_T  72      // Tcy in a horizontal sync pulse (4.7us)
#define BPORCH_T 90      // Tcy in a back porch (4.7us)
#define PIX_T    2       // Tcy in each pixel
#define LINE_T   HRES * PIX_T      // Tcy in each horizontal image line
```

```
// Text Page array
unsigned char  VMap[ COLS * ROWS];
unsigned char *VPtr, *LPtr;

// reordered Font
unsigned char RFont[F_SIZE*8];
unsigned char *FPtr;

volatile int HCount, VCount, RCount, VState, HState;

// next state table
int VS[4] = { SV_SYNC, SV_POSTEQ, SV_LINE, SV_PREEQ};
// next counter table
int VC[4] = { VSYNC_N,  POSTEQ_N,     VRES,   PREEQ_N};
```

The same routines we developed for the TextOnGPage project can now be added directly to this project.

```
void haltVideo()
{
    T3CONbits.TON = 0;   // turn off the vertical state machine
} //haltVideo


void initScreen( void)
{
    int i, j;
    char *v;

    v = VMap;

    // clear the screen
    for ( i=0; i < (ROWS); i++)
        for ( j=0; j < (COLS); j++)
            *v++ = 0;
} //initScreen


int cx, cy;

void putcV( int a)
{
    // check if char in font range
    a -= F_OFFS;
    if ( a < 0)        a = 0;
    if ( a >= F_SIZE)  a = F_SIZE-1;
```

```
    // check page boundaries
    if ( cx >= COLS)           // wrap around x
    {
        cx = 0;
        cy++;
    }
    cy %= ROWS;                // wrap around y

    // find first row in the video map
    VMap[ cy * COLS + cx] = a;

    // increment cursor position
    cx++;
} // putcV


void putsV( unsigned char *s)
{
    while (*s)
        putcV( *s++);
} // putsV


void pcr( void)
{
    cx = 0;
    cy++;
    cy %= ROWS;
} // pcr
```

We can save the new project file as "TextPage.c" and create a new include file "TextPage.h" as well.

```
/*
** TextPage.h
**
** Text Page Video Module
**
*/


#define ROWS    25      // rows of text
#define COLS    48      // columns of text


// Text Page array
extern unsigned char  VMap[ COLS * ROWS];
```

```
// initializes the video output
void initVideo( void);

// stops the video output
void haltVideo();

// clears the video map
void initScreen( void);

// cursor
extern int cx, cy;

void putV( int a);

void putsV( unsigned char *s);

void pcr( void);


#define home()      { cx=0; cy=0;}
#define clrscr()    { initScreen(); home();}
#define AT( x, y)   { cx = (x); cy = (y);}
```

Testing the text page performance

In order to test the new text page video module, we could try to modify an example seen in a previous chapter: the Matrix demo. Back then we were using the asynchronous serial communication module (UART1) to communicate with a VT100 computer terminal (or more likely a PC running the Hyper-Terminal program configured for emulation of the historical DEC terminals VT100 protocol). Now we can replace the putcU routine calls used to send a character to the serial port with putcV calls directed at our video interface.

Let's create a new project called "Matrix2" and let's add all the necessary modules to it including: the rand.c, rand.h, textpage.c, textpage.h and finally a new main module that we will call "matrix2.c" or "the-matrix-reloaded.c" if you prefer.

```
/*
**   The Matrix Reloaded
**
*/

#include <p24fj128ga010.h>
#include "../random/rand.h"
#include "../Text/TextPage.h"

#define COL    40
#define ROW    24
#define DELAY 12000
```

```c
#define pcr() {cx = 0; cy++;}

main()
{
    int v[40];  // vector containing lengh of each string
    int i,j,k;

    // 1. initializations
    T1CON = 0x8030; // TMR1 on, prescale 256, Tcy/2

    initVideo();
    clrscr();       // clear the screen
    randomize( 12); // start the random number sequence

    // 2. init each column lenght
    for( j =0; j<COL; j++)
            v[j] = rand()%ROW;

    // 3. main loop
    while( 1)
    {
        home();

        // 3.1 refresh the screen with random columns
        for( i=0; i<ROW; i++)
        {

            // refresh one row at a time
            for( j=0; j<COL; j++)
            {
                // print a random character down to each column lenght
                if ( i < v[j])
                    putcV( 'A' + (rand()%32));
                else
                    putcV(' ');
            } // for j
            pcr();

        } // for i

            // 3.1.1 delay to slow down the screen update
            TMR1 =0;
            while( TMR1<DELAY);
        // 3.2 randomly increase or reduce each column lenght
        for( j=0; j<COL; j++)
```

```
        {
            switch ( rand()%3)
            {
                case 0: // increase length
                        v[j]++;
                        if (v[j]>ROW)
                            v[j]=ROW;
                        break;

                case 1: // decrease length
                        v[j]--;
                        if (v[j]<1)
                            v[j]=1;
                        break;

                default:// unchanged
                        break;
            } // switch
        } // for

    } // main loop
} // main
```

After saving and building the project, run it on the Explorer16 connected to your video device of choice. You will notice how much faster the screen updates, as the program now has direct access to the video memory and there is no serial connection limiting the information transfers (even as fast as the 115,200-baud connection was in our previous demo project). Also, because now every character placed in the video memory can be retrieved and manipulated in place, new tricks are possible to make the video resemble more closely the movie characteristic and somewhat alien scrolling effect.

Besides the visual impression, though, we are now interested in measuring the actual processor overhead imposed by the new video routines that perform the on-the-fly font translation. For this measurement, the MPLAB SIM software simulator is again our tool of choice. As we did in the previous chapters, we can use one of the PORTA pins (RA0) to signal when we are executing code inside one of the three interrupt service routines:

```
void _ISRFAST _T3Interrupt( void)
{
    _RA0=1;
...
    _RA0=0;
} // T3Interrupt
```

```
void _ISRFAST _OC3Interrupt( void)
{
    _RA0=1;
...
    _RA0=0;
} // OC3Interrupt

void _ISRFAST _OC4Interrupt( void)
{
    _RA0=1;
...
    _RA0=0;
} // OC4Interrupt
```

Remember to add the initialization of the TRISA register inside the `initVideo()` function or the main program to enable the RA0 pin output. Then, add both the RG0 pin (responsible for producing the synchronization pulse) and the RA0 pin to the Logic Analyzer window channels.

Rebuild the project and run it for a short while, just enough to get the first few image lines that represent the worst-case scenario, where the most work is produced by the interrupt service routines.

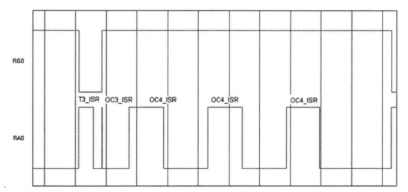

Figure 12-30. Logic analyzer window, measuring the text-page video module overhead.

Using the cursor feature, we can now measure the number of cycles required by each of the four interrupt service routines executed during each of the horizontal line periods. While only the StopWatch tool can give us an exact cycle count, the logic analyzer window can give us a good approximation with a lot less work. My measurements indicate that 384 cycles are spent inside the interrupt service routines of the video module at each 1018 cycles period; that amounts to approximately 38% of the processor available computing power. This is almost double the overhead incurred by the graphic video module routines, but the 20% difference is a price we gladly pay for the great reduction in RAM memory requirements and the increased resolution we gain for all those applications that require a pure text output.

259

Post-flight briefing

In this lesson we have explored the possibility of producing a video output using a minimal hardware interface made up of only three resistors. We learned to use four peripheral modules together to build the complex mechanism required to produce a properly formatted NTSC composite video signal. A 16-bit timer was used to generate the fundamental horizontal synchronization period. Two output compare modules provided intermediate timing references, and finally the SPI module was used in enhanced mode to serialize the video data using the new 8-level-deep by 16-bit FIFO. After developing basic graphic functions to plot individual pixels first and efficiently draw lines, we explored some of the possibilities offered by the availability of a graphic video output, including unidimensional and two-dimensional functions graphing. After briefly exploring the world of fractals, we changed gears to look at the problem of displaying text. First we developed routines to add text to the graphic page, and later we developed a new video module specifically optimized for text display only.

Tips and tricks

The final touch, to complete our brief excursion into the world of graphics, would be to add some animation to our video output libraries. To make the motion fluid and avoid an annoying continuous flicker of the image on the screen, we would need to adopt a technique known as "double buffering." This requires us to have two image buffers in use at any point in time. One is the "active" buffer and its contents are shown on the screen while the other "hidden" buffer is being drawn. When the second buffer drawing is complete, the two are swapped. The first buffer, not visible anymore, is cleared and the drawing process starts again. The only limitation with implementing this technique in our case is represented by the RAM memory size. To make two image buffers fit in the 8 kbytes of memory of the PIC24fj128ga010, while leaving some space for variables and stack, we will need to reduce the image resolution. A pair of image buffers of 160 × 160, for example, would fit as each would require only 3,200 bytes:

```
int _FAR V1Map[VRES * (HRES/16)];
int _FAR V2Map[VRES * (HRES/16)];
```

The only other changes required to the project would be:

1. Replace direct references to the `VMap[]` array with references to pointers

2. Make the interrupt-driven state machine that refreshes the screen use a pointer to the active buffer:

   ```
   int *VA;
   ```

3. Make the plotting and drawing functions use a pointer to the hidden buffer:

   ```
   int *VH;
   ```

The swap between the two buffers can then be performed swapping only two pointers:

```
void swapV( void)
{
    int * V;
    while ( VCount != 1);        // wait until the end of the frame
    V = VA; VA = VH; VH = V;     // at the next VSynch it will swap the screen
} //swapV
```

Notice that care must be taken not to perform the swap in the middle of a frame, but synchronized with the end of a frame and the beginning of the next.

Exercises

1. Replace the `"write.c"` function to redirect the `"stdio.h"` library function output to the text/graphic screen.

2. Add the PS/2 keyboard input support to provide a complete console.

Books

* R. Koster, 2004

   A Theory of Fun for Game Design
   Paraglyph Press

   You must take game design seriously. Or maybe not?

Links

* *http://en.wikipedia.org/wiki/Zx_spectrum*

   The Sinclair ZX Spectrum was one of the first personal computers (home computers as they used to be called) launched in the early 1980s. Its graphic capabilities were very similar to those of the graphic libraries we developed in this project. Although it used several custom logic devices to provide the video output, its processing power was less than a quarter that of the PIC24. Still, the limited ability to produce color (only 16 colors with a resolution of a block of 8 × 8 pixels) enticed many programmers to create thousands of challenging and creative video games.

CHAPTER **13**

Mass storage

In This Chapter

- The SD™/MMC card physical interface
- Interfacing to the Explorer16 board
- Starting a new project
- Selecting the SPI mode of operation

- Sending commands in SPI mode
- Completing the SD/MMC card initialization
- Reading data from an SD/MMC card
- Writing data to an SD/MMC card
- Using the SD/MMC interface module

The relationship between weight (mass) and performance of an airplane is generally well understood by most pilots and nonpilots too. Try to put too much weight on those wings and the takeoff is going to be longer—much longer, or actually so long that there is not enough runway to continue and there is no takeoff at all. Ouch!

The more common problem seems to be in understanding how much all that stuff that you (or your significant other) want to bring along actually weighs. Packing the airplane for a trip with friends or family is just like packing your backpack for an excursion in the outdoors. The fact that everything seemed to fit in does not mean you will be able to lift it. As a pilot you won't be allowed to guess at it; you will have to compile a weight and balance sheet and, if necessary, use a scale to determine the exact numbers and decide what to sacrifice: some of the load or maybe some of the fuel. One thing that I can strongly discourage you from doing, though, is to ask your significant other to step on the scale.

Flight plan

In many embedded-control applications you might find a need for a larger nonvolatile data storage space, well beyond the capabilities of the common serial EEPROM devices we interfaced to in previous chapters and certainly larger than the Flash program memory available inside the microcontroller itself. You might be looking for orders of magnitude more, hundreds of megabytes and possibly gigabytes. If you own a digital camera, an MP3 player or even a cell phone, you have probably become familiar with the storage requirements of consumer multimedia applications and with the available mass-storage technologies. Hard disk drives have become smaller and less power thirsty, but also a multitude of solid-state solutions (based once more on Flash technologies like CompactFlash®, SmartMedia™, Secure Digital (SD), Memory Stick® and others) have flooded the market. Due to the volumes absorbed by the consumer market, the price range has been reduced to a point where it is possible, if not convenient, to integrate these devices into embedded-control applications.

263

In this lesson we will learn how to interface one of the most common and inexpensive mass-storage device types to a PIC24 microcontroller using the smallest amount of processor resources.

The flight

Each one of the many competing mass-storage technologies has its strengths and weaknesses, as each one was designed for a somewhat different target application. We will choose the mass-storage media according to the following criteria:

- Wide availability of the memory and required connectors.

- Small pin count required by the physical interface (serial).

- Large memory capacity.

- Open specifications available.

- Ease of implementation.

- Low cost of the memory and the required connectors.

The Secure Digital (SD) standard compares favorably in all those aspect as it is today one of the most commonly adopted mass-storage media for digital cameras and many other multimedia consumer applications. The SD card specifications represent an evolution of a previous technology known as Multi Media Card, or MMC, with which they are still partially (forward) compatible both electrically and mechanically. The Secure Digital Card Association (SDCA) owns and controls the technical specification standards for the SD memory card and they require that all companies who plan to actively engage in the design, development, manufacture or sale of products that utilize the SD specifications must become members of the association. As of this writing, a general SDCA membership will cost you $2,000 in annual fees. The Multi Media Card Association (MMCA) on the other side does not require implementers to necessarily become members, but makes copies of the MMC specifications available for sale starting at $500. So both technologies are far from free, nor open. Fortunately, there is a subset of the SD specifications that has been released to the public by the SDCA in the form of a "simplified physical specification." This information is all we need to develop a basic understanding of the SD/MMC memory technology and get started designing a PIC24 mass-storage interface.

The SD/MMC card physical interface

SD cards require only nine electrical contacts, and an SD/MMC-compatible connector, which can be purchased on most online catalogs for less than a couple of dollars, requires only a couple of pins more to account for insertion detection and write-protection switch sensing. There are two main modes of communication available: the first one (known as the SD bus) is original to the SD/MMC standard and it requires a nibble (4-bit) wide bus interface; the second mode is serial and is based on the popular SPI bus standard. It is this second mode that makes the SD/MMC mass-storage devices particularly appealing for all embedded-control applications, as most microcontrollers will either have a hardware SPI interface available or will be able to easily emulate one (bit-banging) with a reduced number of I/Os. Finally, the physical specifications of the SD/MMC cards indicate an operating voltage range of 2.0V to 3.6V that is ideally suited for all applications with modern microcontrollers implemented in advanced CMOS processes, as is the case with the PIC24 family.

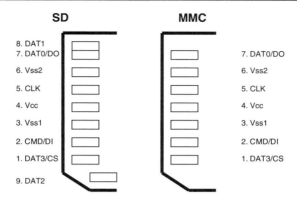

Figure 13-1. SD card and MMC card connectors pin-out.

Interfacing to the Explorer16 board

Unfortunately, although the number of electrical connections required for the SPI interface is very low, all SD/MMC card connectors available on the market are designed for surface-mount applications only, which makes it almost impossible to use the prototyping area of the Explorer16 demonstration board. To facilitate this lesson and the following lessons that will make use of mass-storage devices, complete schematics and PCB layout information for an expansion board have been published on the companion web site *http://www.flyingthePIC24.com*. The expansion board also has interfaces that will be used in the following chapters of the book.

Since in the previous chapter we have used the first SPI peripheral module to produce a video output and the application does not allow for sharing of the resource, we will share instead the second SPI module (SPI2) between the SD card interface and the EEPROM interface using separate Chip Select signals for the two. In addition to the usual SCK, SDI and SDO pins, we will provide pull-ups for the unused pins (reserved for the 4-bit wide SD bus interface) of the SD/MMC connector and for two more pins that will be dedicated to the Card Detect and Write Protect signals.

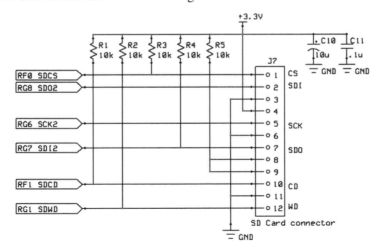

Figure 13-2. SD/MMC card interface to Explorer16 demo board

Starting a new project

After creating a new project (using the usual checklist) we will start by creating the basic initialization routines for all the necessary I/Os and configuring the SPI2 module:

```
/*
** SD card interface
**
*/

#include <p24fj128ga010.h> // pin out definitions

#define SDWD    _RG1    // Write Protect input
#define SDCD    _RF1    // Card Detect input
#define SDCS    _RF0    // Card Select output

void initSD( void)
// initializes the I/Os and peripheral modules (SPI2)
{
    SDCS = 1;               // default Card not-selected (high)
    _TRISF0 = 0;           // make only Card select an output pin

    // init the spi module for a slow (safe) clock speed first
    SPI2CON1 = 0x013c;  // CKE=1, SMP=0, CKP=0, prescale 1:64

    SPI2STAT = 0x8000;  // enable the SPI2 peripheral
}   // initSD
```

In particular, in the SPI2CON1 register we need to configure the SPI module to operate in master mode with the proper clock polarity, clock edge, input sampling point and initial clock frequency. The clock output (SCK) must be enabled and set low when idle. The sampling point for the SDI input must be centered. The frequency is controlled by means of two prescalers (primary and secondary) that divide the main processor cycle clock (Tcy) to generate the SPI clock signal. After power up and until the SD card is properly initialized, we will have to reduce the clock speed to a safe setting (below 400 kHz); therefore we will use the primary prescaler setting 1:64 to obtain a 250-kHz clock signal. This is just a temporary arrangement; after sending only the first few commands, we will be able to speed up the communication considerably.

Notice how only the RF0 pin, controlling the Card Select signal, needs to be manually configured as an output pin, while RG6 and RG8 (corresponding to the pins SCK2 and SDO2) are automatically configured as outputs when we enable the SPI2 peripheral.

Selecting the SPI mode of operation

When an SD/MMC card is inserted in the connector and powered up, it is in the default mode of communication: the SD bus mode. In order to inform the card that we intend to communicate using the alternative SPI mode, all we need to do is to select the card (sending the SDCS pin low) and start sending the first RESET command. We can rest assured that, once in the SPI mode, the card will not be able to change back to the SD bus mode unless the power supply is cycled. This means, though, that if the card is removed from the slot without our knowledge and then reinserted, we will have to make sure that the initialization routine or at least the reset command is repeated in order to get back to the SPI mode. We can detect the card presence at any time by checking the status of the RF1 pin connected to the CD line.

Sending commands in SPI mode

In SPI mode, commands are sent to an SD/MMC card as packets of six bytes, and all responses from the SD card are provided with multiple-byte data blocks of variable length. So all we need to communicate with the memory card is the usual basic SPI routine to send and receive (the two operations are really the same as we have seen in the previous chapter) a byte at a time:

```
// send one byte of data and receive one back at the same time
unsigned char writeSPI( unsigned char b)
{
    SPI2BUF = b;                    // write to buffer for TX
    while( !SPI2STATbits.SPIRBF);   // wait for transfer to complete
    return SPI2BUF;                 // read the received value
}// writeSPI
```

For improved code readability and convenience, we will also define two more macros that will mask the same writeSPI() function as a pure readSPI(), or just as a clock output function clockSPI(). Both macros will send a dummy byte of data (0xFF):

```
#define readSPI()   writeSPI( 0xFF)
#define clockSPI()  writeSPI( 0xFF)
```

To send a command we will start selecting the card (SDCS low) and sending a packet composed of three parts through the SPI port.

BYTE 1	BYTE 2	BYTE 3	BYTE 4	BYTE 5	BYTE 6
7 6 5 4 3 2 1 0	7 6 5 4 3 2 1 0	7 6 5 4 3 2 1 0	7 6 5 4 3 2 1 0	7 6 5 4 3 2 1 0	7 6 5 4 3 2 1 0
COMMAND	ADDRESS				CRC

Figure 13-3. SPI-mode SD/MMC Card command format.

The first part is a single byte containing a command index. The following definitions will cover all the commands we will be using for this project:

```
// SD card commands
#define RESET          0    // a.k.a. GO_IDLE (CMD0)
#define INIT           1    // a.k.a  SEND_OP_COND (CMD1)
#define READ_SINGLE    17   // read a block of data
#define WRITE_SINGLE   24   // write a block of data
```

The command index is followed by a 32-bit memory address. It is an `unsigned long` integer value that must be sent MSB first. For convenience we will define a new type to represent such large address fields, which we will call LBA, borrowing from a term used in other mass-storage applications to represent a very large address to a generic block of data:

```
typedef unsigned long LBA;        // logic block address, 32 bit wide
```

Finally, the command packet is completed by a one-byte CRC. This Cyclic Redundancy Check (CRC) feature is always used in SD bus mode to make sure that every command and every block of data transmitted on the bus is free from errors. But as soon as we switch to the SPI mode (after sending the RESET command) the CRC protection will be automatically disabled, as the card will assume that a direct and reliable connection to the host, the PIC24 in our case, is available. By taking advantage of this default behavior, we can considerably simplify our code replacing the CRC calculation with a precomputed value. This will be the CRC code of the RESET command, and it will be ignored for all the subsequent commands for which the CRC field will be a "don't care." Here is the first part of the sendSDCmd() function:

```
int sendSDCmd( unsigned char c, LBA a)
// sends a 6 byte command block to the card and leaves SDCS active
{
    int i, r;

    // enable SD card
    SDCS = 0;

    // send a comand packet (6 bytes)
    writeSPI( c | 0x40);     // send command + frame bit
    writeSPI( (unsigned char) a>>24);       // msb of the address
    writeSPI( a>>16);
    writeSPI( a>>8);
    writeSPI( a);            // lsb
// NOTE only CMD0-RESET requires an actual CRC (once in SPI mode CRC is disabled)
    writeSPI( 0x95); // send CRC of RESET, for all other cmds it's a don't care
```

After sending all six bytes to the card, we wait in a loop for a response byte (we will in fact keep sending dummy data continuously clocking the SPI port). The response will be 0xFF (basically the SDI line will be kept high) until the card is ready to provide us with a proper response code. The specifications indicate that up to 64 clock pulses (eight bytes) might be necessary before a proper response is received. Should we exceed this limit we would have to assume a major malfunctioning of the card and abort communication.

```
    // now wait for a response up to 8 bytes delay
    i = 9;
    do {
        r = readSPI();       // check if ready
        if ( r != 0xFF) break;
    } while ( --i > 0);

    return ( r);

/* return response
    FF - timeout, no answer
    00 - command accepted
    01 - command received, card in idle state  (after RESET)
    other errors
*/


} // sendSDCmd
```

If we receive a response code, each bit if set will provide us with an indication of a possible problem:

```
bit 0 = Idle state
bit 1 = Erase Reset
bit 2 = Illegal command
bit 3 = Communication CRC error
bit 4 = Erase sequence error
bit 5 = Address error
bit 6 = Parameter error
bit 7 = Always 0
```

Notice that on return the sendSDCmd() function leaves the SD card still selected (SDCS low) so that commands such as Block Write and Block Read, that require additional data to be sent or received from the card, will be able to proceed. In all other commands that do not require additional data transfers, we will have to remember to deselect the card (set SDCS high) immediately after the function call. Furthermore, since we want to share the SPI2 port with other peripherals (for example, the Serial EEPROM mounted on the Explorer16 board) we need to make sure that the SD/MMC card receives a few more clock cycles (8 will suffice) immediately after the rising edge of the chip select line (SDCS). According to the SD/MMC specifications this will allow the card to complete a few important house-keeping chores, including the proper release of the SDO line, essential to allow other devices on the same bus to communicate properly.

Another pair of macros will help us perform this consistently:

```
#define disableSD() SDCS = 1; clockSPI()
#define enableSD()  SDCS = 0
```

Completing the SD/MMC card initialization

Before the card can be effectively used for mass-storage applications, there is a well-defined sequence of commands that needs to be completed. This sequence is defined in the original MMC card specifications and has been modified only slightly by the SD card specifications. Since we are not planning on using any of the advanced features specific to the SD card standard, we will use the basic sequence as defined for MMC cards for maximum compatibility. There are five parts of this sequence:

1. the card is inserted in the connector and powered up.

2. the CS line is initially kept high (card not selected).

3. more than 74 clock pulses must be provided before the card becomes capable of receiving commands.

4. the card must then be selected and a RESET(CMD0) command provided: the card should respond entering the Idle state and (activating the SPI mode).

5. an INIT(CMD1) command is provided and will be repeated until the card exits the Idle state.

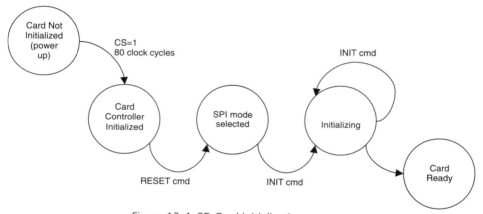

Figure 13-4. SD Card initialization sequence

The following segment of the function initMedia() will perform these initial five steps:

```
int initMedia( void)
{
    int i, r;


    // 1. while the card is not selected
    SDCS = 1;

    // 2. send 80 clock cycles to start up
    for ( i=0; i<10; i++)
        clockSPI();
```

```
// 3. then select the card
SDCS = 0;

// 4. send a reset command to enter SPI mode
r = sendSDCmd( RESET, 0); SDCS = 1;
if ( r != 1)
    return 0x84;

// 5. send repeatedly INIT
i = 10000;              // allow for up to 0.3s before timeout
do {
    r = sendSDCmd( INIT, 0); SDCS = 1;
    if ( r) break;
} while( --i > 0);
if (( i==0) || ( r!=1))
    return 0x85;          // timed out
```

The initialization command can require quite some time, depending on the size and type of memory card, normally measured in several tenths of a second. Since we are operating at 250 kb/s, each byte sent will require 32 μs. Accounting for six bytes for every command retry and using a timeout count of 10,000 will provide us with a generous timeout limit of approximately two seconds.

It is only upon successful completion of this sequence that we will be allowed to finally switch gears and increase the clock speed to the highest possible value supported by our hardware. With minimal experimentation you will find that an Explorer16 board, with a properly designed daughter board providing the SD/MMC connector, can easily sustain a clock rate as high as 8 MHz. This value can be obtained by reconfiguring the SPI primary prescaler for a 1:1 ratio and using the secondary prescaler for a 1:2 ratio. We can now complete the `initMedia()` function with the last segment:

```
// 6. increase speed
SPI2STAT = 0;           // disable momentarily the SPI2 module
SPI2CON1 = 0x013b;      // change prescaler to 1:2
SPI2STAT = 0x8000;      // re-enable the SPI2 module

return 0;

} // init media
```

Reading data from an SD/MMC card

SD/MMC cards are solid-state devices containing typically large arrays of Flash memory, so we would expect to be able to read and write any amount of data (within the card capacity limits) at any desired address. In reality, compatibility considerations with many previous (legacy) mass-storage technologies have imposed a number of constraints on how we can access the memory. In fact all operations are defined in blocks of a fixed size that by default is 512 bytes. It is not a coincidence that 512 bytes is the exact standard size of a data "sector" of a typical personal computer hard disk. Although this can be changed with an appropriate command, we will maintain the default setting so as to take advantage of this compatibility. We will develop a set of routines that will allow us in the following chapter to implement a complete file system compatible with the most common PC operating systems. This way we will be able to access files written on the card by a personal computer and, vice versa a personal computer will be able to access files written by our applications.

The READ_SINGLE (CMD17) is all we need to initiate a transfer of a single "sector" from a given address in memory. The command takes as an argument a 32-bit "byte" address though, so to avoid confusion, in the following sections we will uniformly use only LBAs or block addresses and we will obtain an actual byte address by multiplying the LBA value by 512 just before passing the parameter to the READ_SINGLE command.

We can use the sendSDCmd() function developed above to initiate the read sequence (it will select the card and leave it selected), and after checking the returned response code for errors (there should be none), we will wait for the memory card to send a specific token: DATA_START. This uniquely identifies the beginning of the block of data. Again here, as during the initialization phases, it is important to impose a timeout, although we can be generous. Since only the readSPI() function is called repeatedly (sending/receiving only one byte at a time) while waiting for the data token, a timeout counter of 10,000 will provide an effective time limit of approximately 0.32 seconds (extremely generous).

Once the token is identified, we can confidently read in a rapid sequence all 512 bytes composing the requested block of data. They will be followed by a 16-bit CRC value that we should read, although we will have no use for it.

It is only at this point that we will deselect the memory card and terminate the entire read command sequence.

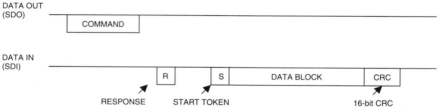

Figure 13-5. Data transfer during a READ_SINGLE command.

The routine `readSECTOR()` performs the entire sequence in a few lines of code.

```
// SD card responses
#define DATA_START                  0xFE

int readSECTOR( LBA a, char *p)
// a        LBA requested
// p        pointer to data buffer
// returns  TRUE if successful
{
    int r, i;

    READ_LED = 1;

    r = sendSDCmd( READ_SINGLE, ( a << 9));
    if ( r == 0)      // check if command was accepted
    {
        // wait for a response
        i = 10000;
        do{
            r = readSPI();
            if ( r == DATA_START) break;
        }while( --i>0);

        // if it did not timeout, read a 512 byte sector of data
        if ( i)
        {
            for( i=0; i<512; i++)
                *p++ = readSPI();

            // ignore CRC
            readSPI();
            readSPI();

        } // data arrived

    } // command accepted

    // remember to disable the card
    disableSD();
    READ_LED = 0;

    return ( r == DATA_START);     // return TRUE if successful
} // readSECTOR
```

To provide a visual indication of activity on the memory card similar to that used on hard drives and floppy disk drives, we have assigned one of the LEDs available on the Explorer16 board as the "read" LED, hoping this will help prevent a user from removing the card while in use. The LED is turned on before each read command and turned off at the end. Other strategies are possible, however. For example, similar to the common practice on USB Flash drives, an LED could be turned on as soon as the card is initialized, regardless of whether an actual command is performed on it at any given point in time. Only calling a deinitialization routine would turn the LED off and indicate to the user that the card can be removed.

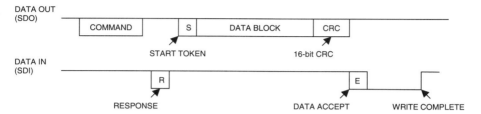

Figure 13-6. Data transfer during a WRITE_SINGLE command.

Writing data to an SD/MMC card

Based on the same consideration we made for the read functions, we will develop a write function that will be similarly constrained to operate on "Sectors"—that is, blocks of 512 bytes of data. The write sequence will use, as you would expect, the WRITE_SINGLE command, but this time the data transfer will be in the opposite direction. Once we make sure that the command is accepted, we will immediately start sending the DATA_START token and right after it all 512 bytes of data, followed by two more bytes for the 16-bit CRC (any dummy value will do). At this point we will pause and check that a new token: DATA_ACCEPT is sent by the card. It will confirm that the entire block of data has been received and the write operation has started. While the card is busy writing it will keep the SDO line low. Waiting for the completion of the write command will require a new loop where we will wait for the SDO line to return high. Once more, a timeout must be imposed to limit the amount of time allowed to the card to complete the operation. Since all SD/MMC memories are based on Flash memory technology, we can expect the time typically required for a write operation to be considerably longer than that required for a read operation. A timeout value of 10,000 would provide us again with a 0.3-s limit that is more than sufficient to accommodate even the slowest memory card on the market.

```
#define DATA_ACCEPT              0x05

int writeSECTOR ( LBA a, char *p)
// a        LBA of sector requested
// p        pointer to sector buffer
// returns  TRUE if successful
{
    unsigned r, i;
    WRITE_LED = 1;

    r = sendSDCmd( WRITE_SINGLE, ( a << 9));
    if ( r == 0)    // check if command was accepted
```

```
    {
        writeSPI( DATA_START);
        for( i=0; i<512; i++)
            writeSPI( *p++);

        // send dummy CRC
        clockSPI();
        clockSPI();

        // check if data accepted
        if ( (r = readSPI() & 0xf) == DATA_ACCEPT)
        {
            for( i=10000; i>0; i--)
            {// wait for end of write operation
                if ( r = readSPI())
                    break;
            }
        } // accepted
        else
            r = FAIL;
    } // command accepted

    // to disable the card and return
    disableSD();
    WRITE_LED = 0;

    return ( r);        // return TRUE if successful

} // writeSECTOR
```

Similarly to the read routine, a second LED has been assigned to indicate when a write operation is being performed and potentially alert the user. If the card is removed during the write sequence, data will most likely be lost.

Save the source we developed so far in a file called "sdmmc.c".

Then add a couple of functions for detecting the card presence and the position of the write protect switch:

```
int detectSD( void)
{
    return ( !SDCD);
} // detect SD

int detectWP( void)
{
    return ( !SDWP);
} // detect WP
```

275

Notice that the WP switch is just providing an indication; it is not connected to a hardware mechanism that would prevent an actual write operation from being performed on the card. It is your responsibility to decide where and when to check for the presence of the WP and to respect it.

Finally, let's create a new include file called "sdmmc.h" to provide the prototypes and basic definitions used in the SD/MMC interface module.

```
/*
** SD/MMC low level card interface
**
*/

#define TRUE     1
#define FALSE    0
#define FAIL     0

// IO definitions
#define READ_LED                  _RA1
#define WRITE_LED                 _RA2


typedef unsigned long LBA;        // logic block address, 32 bit wide


void initSD( void);
int  initMedia( void);

int detectSD( void);
int detectWP( void);

int readSECTOR  ( LBA, char *);
int writeSECTOR ( LBA, char *);
```

Using the SD/MMC interface module

Whether you believe it or not, the six minuscule routines we just developed are all we need to gain access to the seemingly unlimited amount of nonvolatile memory offered by the SD/MMC memory cards. For example, a 512 MB card would provide us with approximately 1,000,000 (yes, that is one million) individually addressable memory blocks (sectors) each 512 bytes large. Note that, as of this writing, SD/MMC cards of this capacity are normally offered for retail in the U.S. for less than $20!

Let's develop a small test program to demonstrate the use of the SD/MMC module. The idea is to simulate a somewhat typical application that is required to save some large amount of data on the SD/MMC memory card. A fixed number of blocks of data will be written in a predetermined range of addresses and then read back to verify the successful completion of the process.

Let's open a new source file and start by adding the usual header and processor-specific include file, followed by the sdmmc.h include file.

```
/*
**  SDMMC read/write Test
**
*/

#include <p24fj128ga010.h>

#include "SDMMC.h"
```

Then, let's define two byte arrays each the size of a default SD/MMC memory block that is 512 bytes.

```
#define B_SIZE          512             // sector/data block size
char    data[ B_SIZE];
char    buffer[ B_SIZE];
```

The test program will fill the first with a specific and easy-to-recognize pattern, and will repeatedly write its contents onto the memory card. The chosen address range will be defined by two constants:

```
#define START_ADDRESS   10000   // start block address
#define N_BLOCKS        1000    // number of blocks
```

The LEDs on PORTA of the Explorer16 demonstration board will provide us with visual feedback about the correct execution of the program and/or any error encountered.

The first few lines of the main program can now be written to initialize the I/Os required by the SD/MMC module and the PORTA pins connected to the row of LEDs.

```
main( void)
{
    LBA addr;
    int i, r;

    // I/O initializations
    TRISA = 0xff00;         // initialize PORTA LEDs output pins
    initSD();               // initialize all I/Os required for the SD/MMC module

    // fill the buffer with "data"
    for( i=0; i<B_SIZE; i++)
        data[i]= i;
```

The next code segment will have to check for the presence of the SD card in the slot/connector. We will wait in a loop for the card detection switch if necessary, and we will provide an additional delay for the contacts to properly debounce.

```
    // wait for card to be inserted
    while( !detectSD());    // assumes SDCD pin is by default an input
    Delayms( 100);          // wait for card contacts debounce and power up
```

We will be generous with the debouncing delay as we want to make sure that the card connection is stable before we start firing "write" commands that could otherwise potentially corrupt other data present on the card. A 100-ms delay is a reasonable delay to use and the Delayms() function can be quickly implemented using any of the PIC24 timers or even the RTCC module. Here is an example that uses the Timer1 timer module and assumes a processor clock of 32 MHz as is the case on the Explorer16 board.

```
void Delayms( unsigned t)
{
    T1CON = 0x8000;      // enable tmr1, Tcy, 1:1
    while (t--)
    {
        TMR1 = 0;
        while (TMR1<16000);
    }
} // Delayms
```

Keeping the debouncing delay function separate from the detectSD() function and the SD/MMC module in general is important, as this will allow different applications to pick and choose the best timing strategy and optimize the resources allocation.

Once we are sure that the card is present, we can proceed with its initialization calling the initMedia() function.

```
// initialize the memory card (returns 0 if successful)
r = initMedia();
if ( r)                  // could not initialize the card
{
    PORTA = r;           // show error code on LEDs
    while( 1);           // halt here
}
```

The function returns an integer value, which is zero for a successful completion of the initialization sequence, or a specific error code otherwise. In our test program, in case of an initialization error we will simply publish the error code on the LEDs and halt the execution, entering an infinite loop. The codes 0x84 and 0x85 will indicate that the initMedia() function steps 4 or 5 have failed, respectively, corresponding to an incorrect execution of the card RESET command and card INIT commands (failure or timeout), respectively.

If all goes well, we will be able to proceed with the actual data-writing phase.

```
else
{
    // fill N_BLOCK blocks/SECTOR with the contents of data buffer
    addr = START_ADDRESS;
    for( i=0; i<N_BLOCKS; i++)
        if (!writeSECTOR( addr+i, data))
        {   // writing failed
            PORTA = 0x0f;
            while( 1);  // halt here
        }
```

The simple for loop performs repeatedly the writeSECTOR() function over the address range from block 10,000 to block 10,999, copying over and over the same data block and verifying at each step that the write command is performed successfully. In case any of the block write commands returns an error, a unique code (0x0f) will be presented on the LEDs and the execution will be halted. In practice this will be equivalent to writing a file of 512,000 bytes.

```
    // verify the contents of each block/SECTOR written
    addr = START_ADDRESS;
    for( i=0; i<N_BLOCKS; i++)
    {   // read back one block at a time
        if (!readSECTOR( addr+i, buffer))
        {   // reading failed
            PORTA = 0xf0;
            while( 1);  // halt here
        }

        // verify each block content
        if ( !memcmp( data, buffer, B_SIZE))
        {   // mismatch
            PORTA = 0xff;
            while( 1); // halt here
        }
    } // for each block
```

Next, we will start a new loop, to read back each data block into the second buffer, and we will compare its contents with the original pattern still available in the first buffer. If the readSECTOR() function should fail we will present an error code (0xf0) on the LEDs display and terminate the test. Otherwise, a standard C library function memcmp() will help us perform a fast comparison of the buffer contents, returning an integer value that is zero if the two buffers are identical as we hope, not zero otherwise. Once more a new unique error indication (0x55) will be provided if the comparison should fail. To gain access to the memcmp() function that belongs to the standard C string library, we will add a new include file to our list:

```
#include <string.h>
```

We can now complete the main program with a final indication of successful execution, lighting up all LEDs on PORTA.

```
    } // else media initialized

    // indicate successful execution
    PORTA = 0xFF;
    // main loop
    while( 1);


} // main
```

If you have added all the required source files: "sdmmc.h", "sdmmc.c" and "sdmmctest.c" to the project, you can now use the standard checklist to build the project and program it on the Explorer16 demonstration board. You will need a daughterboard with the SD/MMC connections, as described at the beginning of the lesson, to actually perform the test. But the effort of building one (or the expense of purchasing one) will be more than compensated for by the joy of seeing the PIC24 perform the test flawlessly in a fraction of a second. The amount of code required was also impressively small.

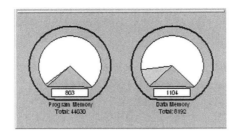

Figure 13-7. MPLAB® Memory Gauges window.

All together, the test program and the SD/MMC access module have used up only 803 words (2409 bytes) of the processor Flash program memory; that is less than 2% of the total memory available. As in all previous lessons, this result was obtained with all compiler optimization options turned off.

Post-flight briefing

In my personal opinion no other mass storage technology is cheaper or easier than this. After all, we can use only a handful of pull-up resistors, a cheap connector, and just a few I/O pins to expand enormously the storage capabilities of our applications. In terms of PIC24 resources required, only the SPI peripheral module has been used and even that could be shared with other applications.

The simplicity of the approach has its obvious limitations though. Data can be written only in blocks of fixed size and its position inside the memory array is completely application specific. In other words, there is no way to share data with a personal computer or other device capable of accessing SD/MMC memory cards unless a "custom" application is developed. Worse, if an attempt is made to use a card already used by a PC, PC data will likely be corrupted and the entire card might require complete re-formatting. In the next lesson we will address these issues by developing a complete file system library.

Tips and tricks

The choice of operating on the default block size of 512 bytes was dictated mostly by historical reasons. By making the low-level access routines in this lesson conform with the standard size adopted by most other mass storage media devices (including hard drives), we made developing the next layer (the file system) easier. But if we were looking for maximum performance, this could have been the wrong choice. In fact, if we were looking for faster write performance, typically the bottleneck of every Flash memory media, we would be better looking at much larger data blocks. Flash memory offers typically very fast access to data (reading) but is relatively slow when it comes to writing. Writing requires two steps: first a large block of data (often referred to as a page) must be erased; then the actual writing can be performed on smaller blocks. The larger the memory array, the larger, proportionally, the erase page size will be. For example, on a 512 MB memory card, the erase page can easily exceed 2 kbytes. While these details are typically hidden from the user, as the main controller inside the card takes care of the erase/write sequencing and buffering, this can have an impact on the overall performance of the application. In fact, if we assume a specific SD card has a 2 kbytes page, writing any amount of data (<2k) would require the internal card controller to perform the following steps:

* Read the contents of an entire 2 kbyte block in an internal buffer.

* Erase it, and wait for the erase-time.

- Replace a portion of the buffer content with the new data.

- Write back the entire 2-kbytes block, and wait for the write-time.

By performing write operations only on blocks of 512 bytes each, to write 2 kbytes of data, our library would have to ask the SD card controller to perform the entire sequence four times, while it could be done in just one sequence by changing the data block length or using a multiple-block write command. While this approach could theoretically increase the writing speed by 400% in the example above, consider the option carefully as the price to pay could be quite high. In fact, consider the following drawbacks:

- The actual memory page size might not be known or guaranteed by the manufacturer, although betting on increasing densities of Flash media (and therefore increasing page size) is pretty safe.

- The size of the RAM buffer to be allocated inside the PIC24 application is increased and this is a precious resource in any embedded application.

- The higher software layers (that we will explore in the next lesson) might be more difficult to integrate if the data block size varies.

- The larger the buffer, the larger the data loss if the card is removed before the buffer is flushed.

Exercises

1. Experiment with various data block sizes to identify where your SD card provides the best write performance. This will give you an indirect indication of the actual page size of the Flash memory device used by the card manufacturer.

2. Experiment with multiple-block write commands by changing the block length to verify how the internal buffering is performed by the SD card controller and if the two methods are equivalent.

Books

- J. Axelson, 2006

  USB Mass Storage: Designing and Programming Devices and Embedded Hosts

  Lakeview Research, WI

  This book continues the excellent series on USB by Jan Axelson. While low-level interfacing directly to a SD/MMC card was easy, as you have seen in this chapter, creating a proper USB interface to a mass storage device is a project of a much higher order of complexity.

Links

- *http://www.mmca.org/home*

  The official web site of the MultiMedia Card Association (MMCA).

- *http://www.sdcard.org/*

  The official web site of the Secure Digital Card Association SDCA.

- *http://www.sdcard.org/sdio/Simplified%20SDIO%20Card%20Specification.pdf*

  The simplified SDIO card specifications. With SDIO, the SD interface is no longer used only for mass storage, but is also a viable interface for a number of advanced peripherals and gizmos, such as GPS receivers, digital cameras and more.

CHAPTER **14**

File I/O

In This Chapter

- ▶ Sectors and Clusters
- ▶ The File Allocation Table (FAT)
- ▶ The root directory
- ▶ The treasure hunt
- ▶ Opening a file
- ▶ Reading data from a file
- ▶ Closing a file

- ▶ Creating a fileio module
- ▶ Testing fopenM() and freadM()
- ▶ Writing data to a file
- ▶ Closing a file, second take
- ▶ Accessory functions
- ▶ Testing the complete fileio module
- ▶ Code size

Every flight during the training should have a precise purpose assigned by the instructor or inspired by the course syllabus used by the school. In each and every lesson, we stated our purpose in a section we called the flight plan, but in aviation an actual flight plan is a different thing. It is a very detailed list containing the times, altitudes, headings, fuel-consumption figures, and so forth for all the segments (legs) composing the flight. For cross-country flights this is an essential tool that will help the pilot stay ahead of the game and be constantly aware of his position and his options in case of emergency. Officially filing the flight plan, calling a Flight Service Station (FSS) and dictating the plan on the phone to a controller, or submitting it via the internet, gives additional advantages. Once the FSS (and ultimately the FAA) knows where, when and along which route you are going, they can keep an eye on you, so to speak. They can track you on their radar (a service called flight following) and, as a minimum, if you are flying too low for them to follow you, they can check that you actually reached your destination at the estimated arrival time or within a reasonable period. If they don't hear from you or if there is no record of your arrival at the destination airport, they will immediately start a search operation. Especially in extreme climates, over mountainous terrain and uninhabited areas, this prompt reaction could be crucial to your life. When it comes to filing flight plans, most pilots have mixed feelings. It feels a bit like when you were a teenager and had to let mom know where you were going to spend the evening; you hate having to do it, although you understand that it is for your own good. Sharing information with mom, I mean the FAA, requires a little effort, but it brings great benefits.

In embedded control, sharing files (information) with a PC can be of great benefit, but you have to know the rules—that is, you need to know how PC file systems work.

Flight plan

In the previous lesson we developed a basic interface module (both software and hardware) to gain access to an SD™/MMC card and support applications that require large amounts of data storage. A similar interface could be built for several other types of mass-storage media, but in this lesson we will rather focus on the algorithms and data structures required to properly share information on the mass-storage device with the most common PC operating systems (DOS, Windows®, and some Linux distributions). In other words, we will develop a module for access to a standard file system known commonly as FAT16. The first FAT file system was created by Bill Gates and Marc McDonald in 1977 for managing disks in Microsoft Disk BASIC. It used techniques that had been available in file systems many years prior and it continued to evolve in numerous versions over the last few decades to accommodate ever larger capacity mass-storage devices and new features. Among the many versions still in use today, the FAT12, FAT16 and FAT32 are the most common ones. FAT16 and FAT32, in particular, are recognized by practically every PC operating system currently in use and the choice between the two is mostly dictated by efficiency considerations and the capacity of the media. Ultimately, for most Flash mass-storage devices of common use in consumer multimedia applications, FAT16 is the file system of choice.

The flight

The name FAT is an acronym that stands for *file allocation table*, which is also the name of one of the most important data structures used in this file system. After all, a file system is just a method for storing and organizing computer files and the data they contain, to make it easy to find and access them. Unfortunately, as often is the case in the history of personal computing, standards and technologies are the fruit of constant evolutionary progress rather than original creation. For this reason, many of the details of the FAT file system that we will reveal in the following sections can only be explained in the context of a struggle to continue and maintain compatibility with an enormous mass of legacy technologies and software over many years.

Sectors and Clusters

Still, the basic ideas at the root of a FAT file system are quite simple. As we have seen in the previous lesson, most mass-storage devices follow a "tradition" derived from the hard-disk technology of managing memory space in blocks of a fixed size of 512 bytes commonly referred to as "sectors." In a FAT file system, a small number of these sectors are reserved and used as a sort of general index: the file allocation table. The remaining (majority) of the sectors are available for proper data storage, but instead of being handled individually, small groups of contiguous sectors are handled jointly to form new, larger entities known as "clusters." Clusters can be as small as one single sector, or can commonly be formed by as many as 64 sectors. It is the use of each cluster and its position that is tracked inside the file allocation table. Therefore, clusters are the true smallest unit of memory allocation in a FAT file system.

The simplified diagram shown in Figure 14-1 illustrates a hypothetical example of a FAT file system formatted for 1022 clusters, each composed of 16 sectors. (Notice that the data area starts with cluster number 2.) In this example each cluster would contain 8 kB of data and the total storage capacity would be about 8 MB.

Note that, the larger clusters are, the fewer are required to manage the entire memory space and the smaller the allocation table required, hence the higher efficiency of the file system. On the contrary, if many small files are to be written, the larger the cluster size, the more space will be wasted. It is typically the responsibility of the operating system, when formatting a storage device for use with a FAT file system, to decide the ideal cluster size to be used for an optimal balance.

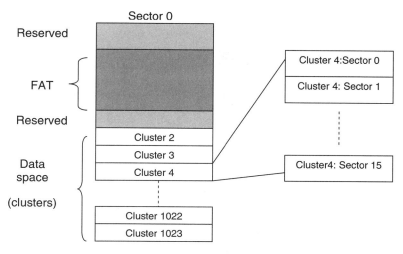

Figure 14-1. Simplified example of a FAT file system layout.

The File Allocation Table (FAT)

In the FAT16 file system, the file allocation table itself contains one 16-bit integer value for each and every cluster. If the cluster is to be considered empty and available, the corresponding position in the table will contain the value 0x0000. If a cluster is in use and it contains an entire file of data, its corresponding position in the table will contain the value 0xFFFF. If a file is larger than the size of a single cluster, a chain of clusters is formed. In the FAT each cluster position in order will contain the number of the following cluster in the chain. The last cluster in the chain will have in the corresponding table position the value 0xFFFF. Additionally, certain unique values are used to mark reserved clusters (0x0001) and bad clusters (0xFFF7). The fact that 0x0000 and 0x0001 have been assigned special meanings is the fundamental reason for the convention of starting the data area with cluster number 2. In the FAT, correspondingly, the first two 16-bit integers are reserved.

In Figure 14-2 you can see an example of the content of a FAT for the system presented in our previous example. Clusters 0 and 1 are reserved. Cluster 2 appears to contain some data, meaning that some or all of the (16) sectors forming the cluster have been filled with data from a file whose size must have been less than 8 kB.

Cluster 3 appears to be the first cluster in a chain of three that also includes cluster 4 and 5. All of cluster 3 and 4 sectors and some or all of cluster 5 sectors must have been filled with data from a file whose size (we can only assume so far) was more than 16 kB but less than 24 kB. All the following clusters appear to be empty and available.

Notice that the size of a FAT itself is dictated by the total number of clusters multiplied by two (two bytes per cluster) and can spread over multiple sectors. In our previous example a FAT of 1024 clusters would have required 2048 bytes, or 4 sectors of 512 bytes each. Also, since the file allocation table is perhaps the most critical structure in the entire FAT file system, multiple copies (typically two) are maintained and allocated one after the other before the beginning of the data space.

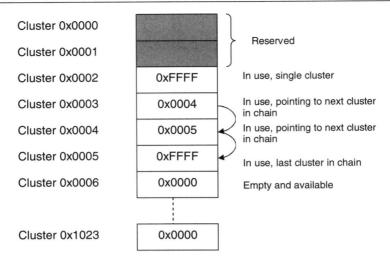

Figure 14-2. Cluster chains in a File Allocation Table.

The Root Directory

The role of the FAT is to keep track of how and where data is allocated. It does not contain any information about the nature of the file to which the data belongs. For that purpose there is another structure called the *root directory*, whose sole purpose is to store file names, sizes, dates, times and a number of other attributes. In a FAT16 file system the root directory (or simply the root from now on) is allocated in a fixed amount of space and a fixed position right between the FAT (second copy) and the first data cluster.

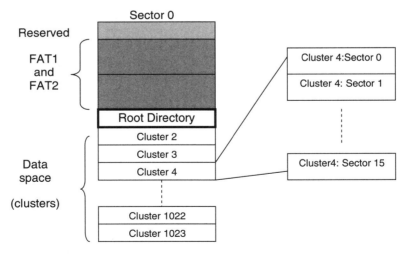

Figure 14-3. Example of a FAT file system layout.

Since both position and size (number of sectors) are fixed, the maximum number of files (or directory entries) in the root directory is limited and determined when formatting the media. Each sector allocated to the root will allow for 16 file entries to be documented, where each entry will require a block of 32 bytes as represented in Figure 14-4.

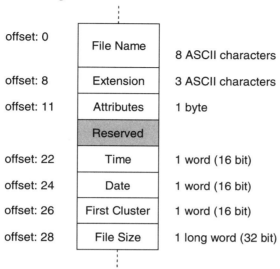

Figure 14-4. Basic Root Directory Entry structure.

The Name and Extension fields are the most obvious, if you are familiar with the older Microsoft operating systems using the 8:3 conventions (the two fields need only to be padded with spaces and the dot can be discarded).

The Attributes field is composed of a group of flags with the meanings shown in Table 14-1:

Bit	Mask	Description
0	0x01	Read Only
1	0x02	Hidden
2	0x04	System
3	0x08	Volume Label
4	0x10	Subdirectory
5	0x20	Archive

Table 14-1. File attributes in a directory entry.

The Time and Date fields (Tables 14-2 and 14-3) refer to the last time the file was modified and must be encoded in a special format to compress all the information in just two 16-bit words.

Bits	Description
15–11	Hours (0–23)
10–5	Minutes (0–59)
4–0	Seconds/2 (0–29)

Table 14-2. Time encoding in a directory entry field.

Bits	Description
15–9	Year (0 = 1980, 127 = 2107)
8–5	Month (1 = January, 12 = December)
4–0	Day (1–31)

Table 14-3. Date encoding in a directory entry field.

Notice how the date field encoding does not allow for the code 0x0000 to be interpreted as a valid date, helping provide clues to the file system when the field is not used or corrupted.

The First Cluster field provides the fundamental link to the FAT. The 16-bit word it contains is nothing but the number of the cluster (could be the only or the first in a chain) containing the file data.

Finally, the Size field, contains in a long integer (32-bit) the size in bytes of the file data.

Looking at the first character of the file name in a directory entry, we can also tell if the entry is currently in use, in which case an actual ASCII-printable character is present, or if the entry is empty, in which case the first byte is a zero and we can also assume that the list of files is terminated as the file system proceeds sequentially using all entries in order. There is a third possibility: when a file is removed from the directory the first character is simply replaced by a special code (0xE5). This indicates that the contents of the entry are no longer valid, and the entry can be reused for a new file at the next opportunity. However, when browsing through the list, searching for a file, we should continue as more active entries might follow.

The treasure hunt

There is much more to say to fully document the structure of a FAT16 file system, but if you have followed the introduction so far, you should now have a reasonable understanding of its core mechanisms and should be ready to dive in for more detail, as we will soon start writing some code.

So far we have maintained a certain level of simplification by ignoring some fundamental questions such as:

* Where do we learn about a storage device capacity?

* How can we tell where the FAT is located?

* How can we tell how many sectors are in each cluster?

* How can we tell where the data space starts?

The answers to all those questions will be found soon by following a sequence of steps that somewhat resemble a child's treasure hunt. We will start using the "sdmmc.c" module functions developed in the previous lesson to initialize the I/Os with the initSD() function first and check for the presence of the card in the slot.

```
// 0. init the I/Os
initSD();

// 1. check if the card is in the slot
if (!detectSD())
{
    FError = FE_NOT_PRESENT;
    return NULL;
}
```

We will proceed then to initialize the storage device with the `initMedia()` function.

```
// 2. initialize the card
if ( initMedia())
{
    FError = FE_CANNOT_INIT;
    return NULL;
}
```

We will also use the standard C libraries (`stdlib.h`) to allocate dynamically two data structures:

```
// 3. allocate space for a MEDIA structure
D = (MEDIA *) malloc( sizeof( MEDIA));
if ( D == NULL)              // report an error
{
    FError = FE_MALLOC_FAILED;
    return NULL;
}

// 4. allocate space for a temp sector buffer
buffer = (unsigned char *) malloc( 512);
if ( buffer == NULL)         // report an error
{
    FError = FE_MALLOC_FAILED;
    free( D);
    return NULL;
}
```

The first one, that will be fully revealed later, is a structure that we will call MEDIA and will be the place where we will collect the answer to all the questions above (perhaps a more appropriate name would have been TREASURE).

The second structure `buffer` is simply a 512-bytes array that will be used to retrieve sectors of data during the hunt.

Notice that to allow the `malloc()` function to successfully allocate memory, you will have to remember to reserve some RAM space for the Heap. Hint: follow the "Project Build" checklist to learn how to reach and modify the linker settings of your project.

Mostly for historical reasons, the first sector (address 0) of each mass storage device will contain what is commonly known as a Master Boot Record (MBR).

Here is how we invoke the `readSECTOR()` function for the first time to access the Master Boot Record:

```
// 5.  get the Master Boot Record
if ( !readSECTOR( 0, buffer))
{
    FError = FE_CANNOT_READ_MBR;
    free( D); free( buffer);
    return NULL;
}
```

A signature, consisting of a specific word value (0x55AA) present in the last word of the MBR sector, will confirm that we have indeed read the correct data.

```
#define FO_SIGN        0x1FE  // MBR signature location (55,AA)


// 6. check if the MBR sector is valid
//      verify the signature word
if (( buffer[ FO_SIGN] != 0x55) ||
    ( buffer[ FO_SIGN +1] != 0xAA))
{
    FError = FE_INVALID_MBR;
    free( D); free( buffer);
    return NULL;
}
```

Once upon a time, this record used to contain actual code to be executed by a PC upon power-up. No personal computer does this anymore, though, and certainly there is no use for that 8086 code for our PIC24 applications. Most of the time, you will find the Master Boot Record sector to be empty, mostly filled with zeros, except for one fixed position starting at offset `0x1BE`. This is where we will find what is called a Partition Table, a table (with only four entries containing 16 bytes each), which has no use on a relatively small memory card like our SD/MMC, but that is kept for compatibility reasons and is identical to the hard-disk partition tables you might have used on your PC. (See Figure 14-5.)

In our applications it is safe to assume that the entire card will have been formatted in a single partition and that this will be the first and only entry (16-byte block) in the table. Of those 16 bytes we will need only a few to deduce the partition size (should include the entire card), the starting sector, and most importantly the type of file system contained. A couple of macros will help us assemble the data from the buffer into words and long words:

```
#define ReadW( a, f) *(unsigned *)(a+f)
#define ReadL( a, f) *(unsigned long *)(a+f)
```

Offset	0	1	2	3	4	5	6	7	8	9	A	B	C	D	E	F	Access ▼	
00000000	00	00	00	00	00	00	00	00	00	00	00	00	00	00	00	00	. . . .	
00000010	00	00	00	00	00	00	00	00	00	00	00	00	00	00	00	00	. . . .	
00000020	00	00	00	00	00	00	00	00	00	00	00	00	00	00	00	00	. . . .	
00000030	00	00	00	00	00	00	00	00	00	00	00	00	00	00	00	00	. . . .	
00000040	00	00	00	00	00	00	00	00	00	00	00	00	00	00	00	00	. . . .	
00000050	00	00	00	00	00	00	00	00	00	00	00	00	00	00	00	00	. . . .	
00000060	00	00	00	00	00	00	00	00	00	00	00	00	00	00	00	00	. . . .	
00000070	00	00	00	00	00	00	00	00	00	00	00	00	00	00	00	00	. . . .	
00000080	00	00	00	00	00	00	00	00	00	00	00	00	00	00	00	00	. . . .	
00000090	00	00	00	00	00	00	00	00	00	00	00	00	00	00	00	00	. . . .	
000000A0	00	00	00	00	00	00	00	00	00	00	00	00	00	00	00	00	. . . .	
000000B0	00	00	00	00	00	00	00	00	00	00	00	00	00	00	00	00	. . . .	
000000C0	00	00	00	00	00	00	00	00	00	00	00	00	00	00	00	00	. . . .	
000000D0	00	00	00	00	00	00	00	00	00	00	00	00	00	00	00	00	. . . .	
000000E0	00	00	00	00	00	00	00	00	00	00	00	00	00	00	00	00	. . . .	
000000F0	00	00	00	00	00	00	00	00	00	00	00	00	00	00	00	00	. . . .	
00000100	00	00	00	00	00	00	00	00	00	00	00	00	00	00	00	00	. . . .	
00000110	00	00	00	00	00	00	00	00	00	00	00	00	00	00	00	00	. . . .	
00000120	00	00	00	00	00	00	00	00	00	00	00	00	00	00	00	00	. . . .	
00000130	00	00	00	00	00	00	00	00	00	00	00	00	00	00	00	00	. . . .	
00000140	00	00	00	00	00	00	00	00	00	00	00	00	00	00	00	00	. . . .	
00000150	00	00	00	00	00	00	00	00	00	00	00	00	00	00	00	00	. . . .	
00000160	00	00	00	00	00	00	00	00	00	00	00	00	00	00	00	00	. . . .	
00000170	00	00	00	00	00	00	00	00	00	00	00	00	00	00	00	00	. . . .	
00000180	00	00	00	00	00	00	00	00	00	00	00	00	00	00	00	00	. . . .	
00000190	00	00	00	00	00	00	00	00	00	00	00	00	00	00	00	00	. . . .	
000001A0	00	00	00	00	00	00	00	00	00	00	00	00	00	00	00	00	. . . .	
000001B0	00	00	00	00	00	00	00	00	00	00	00	00	00	00	00	03	. . . .	
000001C0	35	00	06	08	D8	C1	F1	00	00	00	0F	C9	0E	00	00	00	5 . . . ØÁñ . . . . É	
000001D0	00	00	00	00	00	00	00	00	00	00	00	00	00	00	00	00	. . . .	
000001E0	00	00	00	00	00	00	00	00	00	00	00	00	00	00	00	00	. . . .	
000001F0	00	00	00	00	00	00	00	00	00	00	00	00	00	00	55	AA	. . . . Uª	

Figure 14-5. Hex dump of an MBR sector.

Also the following definitions will point us to the right offset in the MBR.

```
#define FO_FIRST_P    0x1BE  // offset of first partition table
#define FO_FIRST_TYPE 0x1C2  // offset of first partition type
#define FO_FIRST_SECT 0x1C6  // first sector of first partition offset
#define FO_FIRST_SIZE 0x1CA  // number of sectors in partition

    // 7. read the number of sectors in partition
    psize = ReadL( buffer, FO_FIRST_SIZE);
```

```
    // 8. check if the partition type is acceptable
    i = buffer[ FO_FIRST_TYPE];
    switch ( i)
    {
        case 0x04:
        case 0x06:
        case 0x0E:
            // valid FAT16 options
            break;
        default:
            FError = FE_PARTITION_TYPE;
            free( D); free( buffer);
            return NULL;
    } // switch
```

For historical reasons, there are several codes that correspond to a FAT16 file system that we will be able to correctly decode, including 0x04, 0x06 and 0x0E.

Next, we will need to extract the long word (32-bit) value found at offset FO_FIRST_SECT (0x1C6), in the first partition table entry, to proceed in the treasure hunt.

```
    // 9. get the first partition first sector -> Boot Record
    firsts = ReadL( buffer, FO_FIRST_SECT);
```

It contains the address of the next sector that we need to read from the device.

```
    // 10. get the sector loaded (boot record)
    if ( !readSECTOR( firsts, buffer))
    {
        free( D); free( buffer);
        return NULL;
    }
```

It has a signature, similar to the Master Boot Record, located in the last word of the sector, and we need to verify it before proceeding.

```
    // 11. check if the boot record is valid
    //      verify the signature word
    if (( buffer[ FO_SIGN] != 0x55) ||
        ( buffer[ FO_SIGN +1] != 0xAA))
    {
        FError = FE_INVALID_BR;
        free( D); free( buffer);
        return NULL;
    }
```

It is called the (First Partition) Boot Record, and once more it is supposed to contain actual executable code that is of no value to us. (See Figure 14-6.)

Offset	0	1	2	3	4	5	6	7	8	9	A	B	C	D	E	F	Access ▼	
0001E200	EB	00	90	20	20	20	20	20	20	20	20	00	02	20	01	00	ë. ▌	. . . .
0001E210	02	00	02	00	00	F8	77	00	3F	00	10	00	F1	00	00	00	. . . . . øw.?. . .ñ. . .
0001E220	0F	C9	0E	00	80	00	29	13	18	FD	E0	20	20	20	20	20	.É. ▌ .). . ýà
0001E230	20	20	20	20	20	20	46	41	54	31	36	20	20	20	00	00	FAT16 . .
0001E240	00	00	00	00	00	00	00	00	00	00	00	00	00	00	00	00	. . . .
0001E250	00	00	00	00	00	00	00	00	00	00	00	00	00	00	00	00	. . . .
0001E260	00	00	00	00	00	00	00	00	00	00	00	00	00	00	00	00	. . . .
0001E270	00	00	00	00	00	00	00	00	00	00	00	00	00	00	00	00	. . . .
0001E280	00	00	00	00	00	00	00	00	00	00	00	00	00	00	00	00	. . . .
0001E290	00	00	00	00	00	00	00	00	00	00	00	00	00	00	00	00	. . . .
0001E2A0	00	00	00	00	00	00	00	00	00	00	00	00	00	00	00	00	. . . .
0001E2B0	00	00	00	02	00	00	00	00	00	00	00	00	00	00	00	00	. . . .
0001E2C0	00	00	00	00	00	00	00	00	00	00	00	00	00	00	00	00	. . . .
0001E2D0	00	00	00	00	00	00	00	00	00	00	00	00	00	00	00	00	. . . .
0001E2E0	00	00	00	00	00	00	00	00	00	00	00	00	00	00	00	00	. . . .
0001E2F0	00	00	00	00	00	00	00	00	00	00	00	00	00	00	00	00	. . . .
0001E300	00	00	00	00	00	00	00	00	00	00	00	00	00	00	00	00	. . . .
0001E310	00	00	00	00	00	00	00	00	00	00	00	00	00	00	00	00	. . . .
0001E320	00	00	00	00	00	00	00	00	00	00	00	00	00	00	00	00	. . . .
0001E330	00	00	00	00	00	00	00	00	00	00	00	00	00	00	00	00	. . . .
0001E340	00	00	00	00	00	00	00	00	00	00	00	00	00	00	00	00	. . . .
0001E350	00	00	00	00	00	00	00	00	00	00	00	00	00	00	00	00	. . . .
0001E360	00	00	00	00	00	00	00	00	00	00	00	00	00	00	00	00	. . . .
0001E370	00	00	00	00	00	00	00	00	00	00	00	00	00	00	00	00	. . . .
0001E380	00	00	00	00	00	00	00	00	00	00	00	00	00	00	00	00	. . . .
0001E390	00	00	00	00	00	00	00	00	00	00	00	00	00	00	00	00	. . . .
0001E3A0	00	00	00	00	00	00	00	00	00	00	00	00	00	00	00	00	. . . .
0001E3B0	00	00	00	00	00	00	00	00	00	00	00	00	00	00	00	00	. . . .
0001E3C0	00	00	00	00	00	00	00	00	00	00	00	00	00	00	00	00	. . . .
0001E3D0	00	00	00	00	00	00	00	00	00	00	00	00	00	00	00	00	. . . .
0001E3E0	00	00	00	00	00	00	00	00	00	00	00	00	00	00	00	00	. . . .
0001E3F0	00	00	00	00	00	00	00	00	00	00	00	00	00	00	55	AA	. . . . .Uª

Figure 14-6. Hex dump of a Boot Record.

Fortunately, in the same record at fixed and known positions, there are some more of the answers we were looking for and other elements that will help us calculate the rest and complete the map of the entire FAT16 file system. These are the key offsets in the Boot Record buffer:

```
// Partition Boot Record key fields offsets
#define BR_SXC       0xd    // (byte) number of secotrs per cluster
#define BR_RES       0xe    // (word) number of reserved sectors for the boot record
#define BR_FAT_SIZE 0x16    // (word) FAT size in number of sectors
#define BR_FAT_CPY  0x10    // (byte) number of FAT copies
#define BR_MAX_ROOT 0x11    // (odd word) max number of entries in root dir
```

And with the following code we can calculate the size of a cluster:

```
    // 12. determine the size of a cluster
    D->sxc = buffer[ BR_SXC];
    // this will also act as flag that the media is mounted
```

determine the position of the FAT, its size and the number of copies:

```
// 13. determine fat, root and data LBAs
// FAT = first sector in partition (boot record) + reserved records
D->fat = firsts + ReadW( buffer, BR_RES);
D->fatsize = ReadW( buffer, BR_FAT_SIZE);
D->fatcopy = buffer[ BR_FAT_CPY];
```

and find the position of the Root Directory too:

```
// 14. ROOT = FAT + (sectors per FAT *  copies of FAT)
D->root = D->fat + ( D->fatsize * D->fatcopy);
```

Careful now, as we get ready to grab the last few pieces of gold—watch out for a trap!

```
// 15. MAX ROOT is the maximum number of entries in the root directory
D->maxroot = ReadW( buffer, BR_MAX_ROOT) ;
```

Can you see it? No? OK, here is a hint. Look at the value of the `BR_MAX_ROOT` offset as defined a few lines above. You will notice that this is an odd address (`0x11`). This is all it takes for the `ReadW()` macro, which attempts to use it as a word address, to throw a processor trap and reset the PIC24!

We need a special macro (perhaps less efficient) but one that can assemble a word one byte at a time without falling into the trap!

```
// these is the safe versions of ReadW to be used on odd address fields
#define ReadOddW( a, f) (*(a+f) + ( *(a+f+1) << 8))

// 15. MAX ROOT is the maximum number of entries in the root directory
D->maxroot = ReadOddW( buffer, BR_MAX_ROOT) ;
```

The last two pieces of information are easy to grab now. With them, we learn where the data area (divided in clusters) begins and how many clusters are available to our application:

```
// 16. DATA = ROOT + (MAXIMUM ROOT *32 / 512)
D->data = D->root + ( D->maxroot >> 4); // assuming maxroot % 16 == 0!!!

// 17. max clusters in this partition = (tot sectors - sys sectors )/sxc
D->maxcls = (psize - (D->data - firsts)) / D->sxc;
```

It took us as many as 17 careful steps to get to the treasure, but now we have all the information needed to fully figure out the layout of the FAT16 file system present on the SD/MMC memory card (or for that reason almost any other mass-storage media). The treasure, then, is nothing more than another map, a map we will need next to find the files on the mass-storage device. (See Figure 14-7.)

I can now reveal to you the definition of the entire MEDIA structure that we allocated on the heap at the very beginning and we have been so patiently filling. Here is where we will keep the treasure.

```
typedef struct {
    LBA      fat;             // lba of FAT
    LBA      root;            // lba of root directory
    LBA      data;            // lba of the data area
    unsigned maxroot;         // max number of entries in root dir
    unsigned maxcls;          // max number of clusters in partition
    unsigned fatsize;         // number of sectors
    unsigned char fatcopy;    // number of FAT copies
    unsigned char sxc;        // number of sectors per cluster
} MEDIA;
```

We can now assemble all the steps into one essential function that we can call mount() for its similarity to a function available in the Unix family of operating systems.

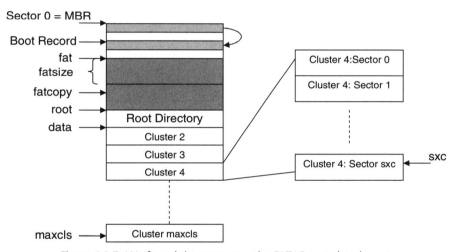

Figure 14-7. We found the treasure—the FAT16 complete layout.

For a mass-storage device to be used in Unix, the file system present on the device must be "mounted," or in other words attached as a new branch of the main (system) file system. Windows users might not be familiar with the concept as they don't have the option to choose if, when, or where the new device file system will be mounted. All new mass-storage devices are automatically and unconditionally mounted by Windows at power-up, or after insertion for removable media, at the very root of the Windows file system by assigning them a single letter identifier ("C:", "D:", "E:", etc.).

```
//-------------------------------------------------------------------
// mount    initializes a MEDIA structure for FILEIO access
//

MEDIA * mount( void)
{
    LBA   psize;      // number of sectors in partition
    LBA   firsts;     // LBA of first sector inside the first partition
    int i;
    unsigned char *buffer;

    ... insert here all 17 steps of our treasure hunt

    // 18. free up the temporary buffer
    free( buffer);
    return D;

} // mount
```

We will also define a global pointer to a MEDIA structure (D) to be used to hold the pointer returned by the mount() function. It will serve as the starting point for the entire file system. Initially, we will assume that only one storage device is available at any given point in time (one connector/slot, one card).

```
// global definitions
MEDIA *D;
```

We will also define an unmount() function that will have the sole duty of releasing the space allocated for the MEDIA structure.

```
//-------------------------------------------------------------------
// unmount    releases the space allocated for the MEDIA structure
//

void unmount( void)
{
    free( D);
    D = NULL;
} // unmount
```

Opening a file

Now that we have figured out the map of the storage device, we can start pursuing our original objective: accessing individual files. In practice, what we will develop in the following parts of this lesson is a set of high-level functions similar to those found in most operating systems for file manipulation. We will need a function to find a file location on the storage device, one for sequentially reading the data from the file and possibly one more to write data and create new files.

In a logical order, we will start developing what we will call the fopenM() function. Its role will be that of finding all possible information regarding a file (if present) and gathering it in a new structure that we will call MFILE. The name of this structure was chosen so as to avoid conflicts with similar structures and functions defined inside the standard C library "stdio.h".

```
typedef struct {
    MEDIA  * mda;              // MEDIA structure
    unsigned char * buffer;    // sector buffer
    unsigned cluster;          // first cluster
    unsigned ccls;             // current cluster in file
    unsigned sec;              // sector in current cluster
    unsigned pos;              // position in current sector
    unsigned top;              // number of data bytes in the buffer
    long     seek;             // position in the file
    long     size;             // file size
    unsigned time;             // last update time
    unsigned date;             // last update date
    char     name[11];         // file name
    char     chk;              // MFILE structure checksum = ~( entry + name[0])
    unsigned entry;            // entry position in cur directory
    char     mode;             // mode 'r', 'w'
    } MFILE;
```

I know at first sight it looks like a lot—it is more than 40 bytes large—but as you will see in the rest of the lesson, we will end up needing all of them. You will have to trust me for now.

Mimicking standard C library implementations (common to many operating systems) the fopenM() function will receive two (ASCII) string parameters: the file name and a "mode" string, containing "r" or "w", that will indicate if the file is supposed to be opened for reading or writing.

```
MFILE *fopenM( const char *filename, const char *mode)
{
    char c;
    int i, r;
    unsigned char *b;          // newly allocated buffer
    MFILE *fp;                 // pointer to newly allocated MFILE structure
    MEDIA *mda=D;              // pointer to MEDIA structure
```

To optimize memory usage an MFILE structure is allocated only when necessary; it is in fact one of the first tasks of the fopenM() function, and a pointer to the data structure is its return value. In case fopenM() should fail, a NULL pointer will act as an error report.

Of course, a prerequisite for opening a file is to have the storage device file system mapped out, and that is supposed to have already been performed by the mount() function. A pointer to a MEDIA structure must have already been deposited in the global D pointer.

```
// 1.  check if a storage device is mounted
if ( D == NULL)        // unmounted
{
    FError = FE_MEDIA_NOT_MNTD;
    return NULL;
}
```

Chapter 14

Since all activity with the storage device must be performed in blocks of 512 bytes, we will need that much space to be allocated for it to act as a read/write buffer.

```
// 2. allocate a buffer for the file
b = (unsigned char*)malloc( 512);
if ( b == NULL)
{
    FError = FE_MALLOC_FAILED;
    return NULL;
}
```

Only if that amount of memory is available can we proceed and allocate some more memory for the MFILE structure proper.

```
// 3. allocate a MFILE structure on the heap
fp = (MFILE *) malloc( sizeof( MFILE));
if ( fp == NULL)                // report an error
{
    FError = FE_MALLOC_FAILED;
    free( b);
    return NULL;
}
```

The buffer pointer and the MEDIA pointers can now be recorded inside the MFILE data structure.

```
// 4. set pointers to the MEDIA structure and buffer
fp->mda = D;
fp->buffer = b;
```

The file name parameter must be extracted; each character must be translated to upper case (using the standard C library functions defined in "ctype.h") and padded, if necessary, with spaces to an eight-character length.

```
// 5. format the filename into name
for( i=0; i<8; i++)
{
    c = toupper( *filename++);      // read a char and convert to upper case
    if (( c == '.') || ( c == '\0'))// extension or short name noextension
        break;
    else
        fp->name[i] = c;
} // for
// if short fill the rest up to 8 with spaces
while ( i<8) fp->name[i++] = ' ';
```

Similarly, after discarding the dot, an extension of up to three characters must be formatted and padded.

```
// 6. if there is an extension
if ( c != '\0')
{
    for( i=8; i<11; i++)
    {
        c = toupper( *filename++);      // read char, convert to upper case
        if ( c == '.')
            c = toupper( *filename++);
        if ( c == '\0')                 // short extension
            break;
        else
            fp->name[i] = c;
    } // for
    // if short fill the rest up to 3 with spaces
    while ( i<11) fp->name[i++] = ' ';
} // if
```

While most C libraries provide extensive support for multiple "modes" of access to files, like distinguishing between text and binary files and offering an "append" option, we will accept (at least initially) a subset consisting of two basic options only: "r" or "w".

```
// 7. copy the file mode character  (r, w)
if ((*mode == 'r')||(*mode == 'w'))
    fp->mode = *mode;
else
{
    FError = FE_INVALID_MODE;
    goto ExitOpen;
}
```

With the file name properly formatted, we can now start searching the root directory of the storage device for an entry of the same name.

```
// 8. Search for the file in current directory
if ( ( r = findDIR( fp)) == FAIL)
{
    FError = FE_FIND_ERROR;
    goto ExitOpen;
}
```

Let's leave the details of the search out for now and let's trust the new findDIR() function to return to us one of three possible values: FAIL, NOT_FOUND and eventually FOUND. A possible failure must always be taken into account. After all, before we consider the possibility of major fatal failures of the storage device, there is always the possibility that the user simply removed the card from its slot without our knowledge. If that is the case, as in all prior error cases, we have no business continuing in the process. We better immediately release the memory allocated thus far and return with a NULL pointer after leaving an error code in the dedicated "mail box" FError, just like we did during the mount process.

But if the search for the file is completed without errors (whether it was found or not), we can continue initializing the MFILE structure.

```
// 9. init all counters to the beginning of the file
fp->seek = 0;            // first byte in file
fp->sec = 0;             // first sector in the cluster
fp->pos = 0;             // first byte in sector/cluster
```

The counter seek will be used to keep track of our position inside the file as we will access its contents sequentially. Its value will be a long integer (unsigned long) between 0 and the size of the entire file expressed in bytes. The sec field will keep track of which sector (inside the current cluster) we are currently operating on. Its value will be an integer between 0 and sxc-1, the number of sectors composing each data cluster. Finally, pos will keep track of which byte (inside the current buffer) we are going to access next. Its value will be an integer between 0 and 511.

```
// 10. depending on the mode (read or write)
if ( fp->mode == 'r')
{
```

At this point, different things need to be done depending on whether an existing file needs to be opened for reading or a new file needs to be created for writing. Initially we will complete all the necessary steps for the fopenM() function when invoked in the read ("r") mode, in which case the file better be found.

```
// 10.1 'r' open for reading
if  ( r == NOT_FOUND)
{
    FError = FE_FILE_NOT_FOUND;
    goto ExitOpen;
}
```

If it was indeed found, the findDIR() function will have filled for us a couple more fields of the MFILE structure, including:

- entry: indicating the position in the root directory where the file was found

- cluster: indicating the number of the first data cluster used to store the file data as retrieved from the directory entry

- size: indicating the number of bytes composing the entire file

- time and date of creation

- the file attributes

The first cluster number will become our current cluster: `ccls`.

```
else
{   // found

    // 10.2 set the current cluster pointer on the first file cluster
        fp->ccls = fp->cluster;
```

We now have all the information required to identify the first sector of data into the buffer. The function `readDATA()` (that we will describe in detail shortly) will perform the simple calculation required to convert the `ccls` and `sec` values into an absolute sector number inside the data area and will use the low-level `readSECTOR()` function to retrieve the data from the storage device.

```
    // 10.3 read a sector of data from the file
        if ( !readDATA( fp))
        {
            goto ExitOpen;
        }
```

Notice that the file length is not constrained to be a multiple of a sector size. So it is perfectly possible that only a part of the data retrieved in the buffer belongs to the actual file. The `MFILE` structure field `top` will help us keep track of where the actual file data ends and possibly padding begins.

```
    // 10.4 determine how much data is really inside the buffer
        if ( fp->size-fp->seek < 512)

            fp->top = fp->size - fp->seek;

        else

            fp->top = 512;

    } // found
} // 'r'
```

As this is all we really need to complete the `fopenM()` function (when opening a file for reading), we could return now with the precious pointer to the `MFILE` structure. As an additional safety measure, though, to help flag possible future mistakes related to the use and reuse of pointers, we will compute a simple checksum field that will result correctly only if the entire open function was completed successfully.

```
    // 12.  compute the MFILE structure checksum
    fp->chk = ~( fp->entry + fp->name[0]);

    return fp;
```

Note: Shortly we will be inserting some more code before this point, so don't worry for now about the numbering sequence.

In case any of the previous steps failed, we will exit the function, returning a NULL pointer after having released both the memory allocated for the sector buffer and the MFILE structure.

```
    // 13. Exit with error
ExitOpen:
    free( fp->buffer);
    free( fp);
    return NULL;

} // fopenM
```

In a top-down fashion, we can now complete the two accessory functions used during the development of fopenM(), starting with readDATA():

```
unsigned readDATA( MFILE *fp)
{
    LBA l;

    // calculate lba of cluster/sector
    l = fp->mda->data + (LBA)(fp->ccls-2) * fp->mda->sxc + fp->sec;

    return( readSECTOR( l, fp->buffer))

} // readDATA
```

Notice how we need data and sxc from the MEDIA structure to compute the correct sector number. Very simple!

Similarly, we can create a function to read from the root directory a block of data containing a given entry.

```
unsigned readDIR( MFILE *fp, unsigned e)
// loads current entry sector in file buffer
// returns      FAIL/TRUE
{
    LBA l;

    // load the root sector containing the DIR entry "e"
    l = fp->mda->root + (e >> 4);

    return ( readSECTOR( l, fp->buffer));

} // readDIR
```

We know that each directory entry is 32 bytes long; therefore each sector will contain 16 entries.

The findDIR() function can now be quickly coded as a short sequence of steps enclosed in a search loop through all the available entries in the root directory.

```
unsigned findDIR( MFILE *fp)
// fp        file structure
// return    found/not_found/fail
{
    unsigned eCount;                 // current entry counter
    unsigned e;                      // current entry offset in buffer
    int i, a, c, d;
    MEDIA *mda = fp->mda;

    // 1. start from the first entry
    eCount = 0;
    // load the first sector of root
    if ( !readDIR( fp, eCount))
        return FAIL;
```

We start loading the first root sector, containing the first 16 entries, in the buffer. For each entry we compute its offset inside the buffer:

```
// 2. loop until you reach the end or find the file
while ( 1)
{
// 2.0 determine the offset in current buffer
    e = (eCount & 0xf) * DIR_ESIZE;
```

and we inspect the first character of the entry file name:

```
// 2.1 read the first char of the file name
    a = fp->buffer[ e + DIR_NAME];
```

If its value is 0, indicating an empty entry and the end of the list, we can immediately exit reporting the file name was not found.

```
// 2.2 terminate if it is empty (end of the list)
    if ( a == DIR_EMPTY)
    {
        return NOT_FOUND;
    } // empty entry
```

The other possibility is that the entry was marked as deleted, in which case we will skip it.

```
// 2.3 skip erased entries if looking for a match
    if ( a != DIR_DEL)
    {
```

Otherwise, it's a valid healthy entry, and we should check the attributes to determine if it corresponds to a proper file or any other type of object. The possibilities include: subdirectories, volume labels and long file names. None of them is of our concern, as we will choose to keep things simple and we will steer clear of the most advanced and sometimes patented features of the more recent versions of the FAT file system standard.

```
// 2.3.1 if not VOLume or DIR compare the names
    a = fp->buffer[ e + DIR_ATTRIB];

    if ( !((a & ATT_DIR) || ( a & ATT_VOL)) )
    {
```

We will then compare the file names character by character, looking for a complete match.

```
// compare file name and extension
for (i=DIR_NAME; i<DIR_ATTRIB; i++)
{
    if ( ( fp->buffer[ e + i]) != ( fp->name[i]))
        break;  // difference found
}
```

Only if every character matches, we will extract the essential pieces of information from the entry and we will copy them into the MFILE structure, returning a FOUND code.

```
if ( i == DIR_ATTRIB)
{
    // entry found, fill the mfile structure
    fp->entry = eCount;         // store entry index
    fp->time = ReadW( fp->buffer, e + DIR_TIME);
    fp->date = ReadW( fp->buffer, e + DIR_DATE);
    fp->size = ReadL( fp->buffer, e + DIR_SIZE);
    fp->cluster = ReadL( fp->buffer, e + DIR_CLST);
    return FOUND;
}
    } // not a dir nor a vol
} // not deleted
```

Should the file name and extension differ, we will simply continue our search with the next entry, remembering to load the next sector from the root directory after each group of 16 entries.

```
// 2.4 get the next entry
    eCount++;
    if ( eCount & 0xf == 0)
    {   // load a new sector from the Dir
        if ( !readDIR( fp, eCount))
            return FAIL;
    }
```

We know the maximum number of entries in the root directory (`maxroot`) and we need to terminate our search if we reach the end of the directory without a match indicating `NOT_FOUND`.

```
    // 2.5. exit the loop if reached the end or error
        if ( eCount >= mda->maxroot)
            return NOT_FOUND;                    // last entry reached


    }// while


} // findDIR
```

Reading data from a file

Finally, this is the moment we have waited for so long. The file system is mounted, a file is found and opened for reading, and it is time to develop the `freadM()` function, to read freely blocks of data from it.

```
    unsigned freadM( void * dest, unsigned size, MFILE *fp)
    // fp        pointer to MFILE structure
    // dest      pointer to destination buffer
    // count     number of bytes to transfer
    // returns   number of bytes actually transferred
    {
        MEDIA * mda = fp->mda;  // media structure
        unsigned count=size;    // counts the number of bytes to be transferred
        unsigned len;
```

The name, number and sequence of parameters passed to this function are again supposed to mimic closely those of similarly named functions available in the standard C libraries.

A destination buffer is supplied where the data read from the file will be copied, and a number of bytes is requested while passing the usual pointer to an open `MFILE` structure.

The `freadM()` function will do its best to read as many of the bytes requested as possible from the file, and will return an unsigned integer value to report how many it effectively managed to get. In our simple implementation, if the number returned is not identical to what was requested by the calling application, we will have to assume that something major has happened. The end of file has been reached, most probably, but we will not make a distinction if, instead, another type of failure has occurred—for example, the card has been removed during the process.

As usual, we will not trust the pointer passed in the argument and we will check instead if it is pointing to a valid `MFILE` structure by recalculating and comparing the simple checksum performed by the open function when successfully opening a file.

```
    // 1. check if fp points to a valid open file structure
    if (( fp->entry + fp->name[0] != ~fp->chk ) || ( fp->mode != 'r'))
    {   // checksum fails or not open in read mode
        FError = FE_INVALID_FILE;
        return size-count;
    }
```

Only then we will enter a loop to start transferring the data from the sector data buffer.

```
// 2. loop to transfer the data
while ( count>0)
{
```

Inside the loop, the first condition to check will be our current position, with regard to the total file size.

```
// 2.1 check if EOF reached
if ( fp->seek >= fp->size)
{
    FError = FE_EOF;  // reached the end
    break;
}
```

Notice that this error will be generated only if the application calling the freadM() function will ignore the previous symptom: if the last freadM() call returned with a number of data bytes inferior to what was requested, or if the calling application has requested the exact number of bytes available in the file with the previous calls.

Otherwise we will verify if the current buffer of data has already been used up completely.

```
// 2.2 load a new sector if necessary
if (fp->pos == fp->top)
{
```

If necessary, we will reset our buffer pointers and attempt to load the next sector from the file:

```
fp->pos = 0;
fp->sec++;
```

If we had already used up all the sectors in the current cluster, this might force us to step into the next cluster by peeking inside the FAT and following the chain of clusters:

```
// 2.2.1 get a new cluster if necessary
if ( fp->sec == mda->sxc)
{
    fp->sec = 0;
    if ( !nextFAT( fp, 1))
    {
        break;
    }
}
```

In either case, we load the new sector of data in the buffer, paying attention to verify the possibility that it might be the last one of the file and it might be only partially filled:

```
// 2.2.2 load a sector of data
if ( !readDATA( fp))
{
    break;
}
// 2.2.3 determine how much data is really inside the buffer
if ( fp->size-fp->seek < 512)
    fp->top = fp->size - fp->seek;
else
    fp->top = 512;
} //  load new sector
```

Now that we know we have data in the buffer, ready to be transferred, we can determine how much of it we can transfer in a single chunk:

```
// 2.3 copy as many bytes as possible in a single chunk
// take as much as fits in the current sector
if ( fp->pos+count < fp->top)
    len = count;              // fits all in current sector
else
    len = fp->top - fp->pos;   // take a first chunk, there is more

memcpy( dest, fp->buffer + fp->pos, len);
```

Using the memcpy() function from the standard C libraries (string.h) to move a block of data from the file buffer to the destination buffer, we get the best performance, as these routines are optimized for speed of execution. The pointers and counters can be updated and the loop can be repeated until all the data requested has been transferred.

```
// 2.4 update all counters and pointers
count-= len;              // compute what is left
dest += len;              // advance destination pointer
fp->pos += len;           // advance the pointer in current sector
fp->seek += len;          // advance the seek pointer

} // while count
```

Finally, we can exit the function and return the number of actual bytes transferred in the loop:

```
// 3. return number of bytes actually transferred
return size-count;

} // freadM
```

Closing a file

Since we can only open a file for reading (with the `fopenM()` function as defined so far), there is not much work to perform upon closing the file. We can consider invalidating the checksum created by `fopenM()` but we must remember to free all the memory allocated for the `MFILE` structure and the sector buffer.

```c
unsigned fcloseM( MFILE *fp)
{
    // 1. invalidate the file structure
    fp->chk = fp->entry + fp->name[0];      // set checksum invalid!

    // 2. free up the buffer and the MFILE struct
    free( fp->buffer);
    free( fp);

} // fcloseM
```

Creating the fileio module

We can create a small library module by saving all the functions written so far in a file called "`fileio.c`". We will need to add the usual header with a few include files:

```c
/*
** FILE I/O interface
**
** module: fileio.c
**
*/

// standard C libraries used
#include <stdlib.h>      // NULL, malloc, free...
#include <ctype.h>       // toupper...
#include <string.h>      // memcpy...

#include "sdmmc.h"       // sd/mmc card interface
#include "fileio.h"      // file I/O routines
```

And of course, we will need to create the "`fileio.h`" file too, with all the definitions and prototypes that we wish to publish for future applications to use.

```c
/*
** FILE I/O interface
**
** FAT16 support
**
** module:    fileio.h
*/
```

```
extern char FError;    // mailbox for error reporting

// FILEIO ERROR CODES
#define FE_IDE_ERROR         1    // IDE command execution error
#define FE_NOT_PRESENT       2    // CARD not present
#define FE_PARTITION_TYPE    3    // WRONG partition type, not FAT12
#define FE_INVALID_MBR       4    // MBR sector invalid signature
#define FE_INVALID_BR        5    // Boot Record invalid signature
#define FE_MEDIA_NOT_MNTD    6    // Media not mounted
#define FE_FILE_NOT_FOUND    7    // File not found in open for read
#define FE_INVALID_FILE      8    // File not open
#define FE_FAT_EOF           9    // Fat attempt to read beyond EOF
#define FE_EOF              10    // Reached the end of file
#define FE_INVALID_CLUSTER  11    // Invalid cluster value > maxcls
#define FE_DIR_FULL         12    // All root dir entry are taken
#define FE_MEDIA_FULL       13    // All clusters in partition are taken
#define FE_FILE_OVERWRITE   14    // A file with same name exists already
#define FE_CANNOT_INIT      15    // Cannot init the CARD
#define FE_CANNOT_READ_MBR  16    // Cannot read the MBR
#define FE_MALLOC_FAILED    17    // Malloc could not allocate the MFILE struct
#define FE_INVALID_MODE     18    // Mode was not r.w.
#define FE_FIND_ERROR       19    // Failure during FILE search

typedef struct {
    LBA      fat;            // lba of FAT
    LBA      root;           // lba of root directory
    LBA      data;           // lba of the data area
    unsigned maxroot;        // max number of entries in root dir
    unsigned maxcls;         // max number of clusters in partition
    unsigned fatsize;        // number of sectors
    unsigned char fatcopy;   // number of copies
    unsigned char sxc;       // number of sectors per cluster (!=0 flags media mounted)
    } MEDIA;


typedef struct {
    MEDIA * mda;             // media structure pointer
    unsigned char * buffer;  // sector buffer
    unsigned cluster;        // first cluster
    unsigned ccls;           // current cluster in file
    unsigned sec;            // sector in current cluster
    unsigned pos;            // position in current sector
    unsigned top;            // number of data bytes in the buffer
    long     seek;           // position in the file
    long     size;           // file size
    unsigned time;           // last update time
```

```
        unsigned date;              // last update date
        char    name[11];           // file name
        char    chk;                // checksum = ~( entry + name[0])
        unsigned entry;             // entry position in cur directory
        char    mode;               // mode 'r', 'w', 'a'
        } MFILE;


// file attributes
#define ATT_RO      1       // attribute read only
#define ATT_HIDE    2       // attribute hidden
#define ATT_SYS     4       //  "          system file
#define ATT_VOL     8       //  "          volume label
#define ATT_DIR     0x10    //  "          sub-directory
#define ATT_ARC     0x20    //  "          (to) archive
#define ATT_LFN     0x0f    // mask for Long File Name records


#define FOUND       2       // directory entry match
#define NOT_FOUND   1       // directory entry not found



// macros to extract words and longs from a byte array
// watch out, a processor trap will be generated if the address is not word
aligned
#define ReadW( a, f) *(unsigned *)(a+f)
#define ReadL( a, f) *(unsigned long *)(a+f)

// this is a "safe" version of ReadW to be used on odd address fields
#define ReadOddW( a, f) (*(a+f) + ( *(a+f+1) << 8))



// prototypes
unsigned nextFAT( MFILE * fp, unsigned n);
unsigned newFAT(  MFILE * fp);

unsigned readDIR( MFILE *fp, unsigned entry);
unsigned findDIR( MFILE *fp);
unsigned newDIR ( MFILE *fp);

MEDIA * mount( void);
void    unmount( void);

MFILE *  fopenM  ( const char *name, const char *mode);
unsigned freadM  ( void * dest, unsigned count, MFILE *);
unsigned fwriteM ( void * src, unsigned count, MFILE *);
unsigned fcloseM ( MFILE *fp);
```

Don't worry now if we have not yet fleshed out all the functions; we will continue working on them as we proceed through the rest of the lesson.

Testing fopenM() and freadM()

It might seem like a long time since we built a project for the last time. To verify the code that we have developed so far, we had to reach a critical mass, a minimal core of routines without which no application could have worked. Now that we have this core functionality, we can develop for the first time a small test program to read a file from an SD/MMC card that was formatted with the FAT16 file system.

The idea is to copy a text file (any text file would work) onto the SD/MMC card from your PC, and then have the PIC24, with the new "fileio.c" module, read the file and send its content to the serial port back to the PC (running Hyper terminal or any other terminal or printer available with an RS232 serial port).

This is the main module that you will save as "ReadTest.c".

```
/*
**       ReadTest.c
**
*/

#include <p24fj128ga010.h>

#include "SDMMC.h"
#include "fileio.h"
#include "../delay/delay.h"
#include "../3 comm/conu2.h"


#define B_SIZE   10
char data[ B_SIZE];

main( void)
{
    MFILE *fs;
    unsigned i, r;

    //initializations
    initU2();               //115,200 baud 8,n,1

    putsU2( "init");

    while( !detectSD());    // assumes SDCD pin is by default an input
    Delayms( 100);          // wait for card to power up

    putsU2("media detected");
```

```
if ( mount())
{
    putsU2( "mount");
    if ( fs = fopenM( "name.txt", "r"))
    {
        putsU2("file opened");
        do{
            r = freadM( data, B_SIZE, fs);
            for( i=0; i<r; i++)
                putU2( data[i]);
        } while( r==B_SIZE);
        fcloseM( fs);
        putsU2("file closed");
    }
    else
        putsU2("could not open file");

    unmount();
    putsU2("media unmounted");
}

// main loop
while( 1);
} // main
```

We will use the serial communication module "conu2.c" developed in one of the early lessons and a delay module that will provide a delayms() function similar, to the one we used to test the "sdmmc.c" module in the previous lesson. The sequence of operation is also similar, only this time instead of calling the initMedia() function and then reading and writing directly to sectors of the SD/MMC card, we will call the mount() function to access the FAT16 file system on the card. We will open the data file using its "proper" name, and we will read data from it in blocks of arbitrary length (B_SIZE) and we will send its contents to the serial port of the Explorer16 board.

Once we have exhausted the content of the entire file, we will close the file, deallocating all the memory used.

After creating a new project we will need to add all the necessary modules to the project window including:

- "sdmmc.c"

- "fileio.c"

- "conu2.c"

- "delay.c"

- "readtest.c"

and all the corresponding include files (.h).

Remember to follow the checklists for a new project and for the ICD2 debugger, so that you will remember to set the ICD2 option for the linker; in the same configuration dialog box, remember to add some space for the heap so that we will be able to allocate memory dynamically for the file system structures and buffers (even if 580 bytes should suffice, give the heap ample room to maneuver).

After building the project and programming the executable on the Explorer16 board, we are ready to run the test.

If all goes well, you will be able to see the contents of the text file scrolling on the screen of your terminal of choice, probably too fast for you to read it, except for the last part.

Notice that you can recompile the project and run the test with different sizes for the data buffer from 1 byte to as large as the memory of the PIC24 will allow. The freadM() function will take care of reading as many sectors of data required to fulfill your request as long as there is data in the file.

Writing data to a file

We are far from finished. The "fileio.c" module is not complete until we include the ability to create new files. This requires us to create an fwriteM() function but also to complete a piece of the fopenM() function. So far, in fact, we had fopenM() return with an error code when a file could not be found in the root directory or the mode was not "r". But this is exactly what we want when we open a new file for writing. When we check for the mode parameter value, we now need to add a new option. This time, it is when the file is NOT_FOUND during the first scan of the directory that we want to proceed.

```
else  // 11.  open for 'write'
  {
     if  ( r == NOT_FOUND)
     {
```

A new file needs a new cluster to be allocated to contain its data. The function newFAT() will be used to search in the FAT for an available spot, a cluster that is still marked (with 0x0000) as available. This search could fail and the function could return an error that, among other things, could indicate that the storage device is full and all data clusters are taken. Should the search be successful, though, we will take note of the new cluster position and update the MFILE structure, making it the first cluster of our new file.

```
                    // 11.1 allocate a first cluster to it
                    fp->ccls = 0;                          // indicate brand new file
                    if ( newFAT( fp) == FAIL)
                    { // must be media full
                        FError = FE_MEDIA_FULL;
                        goto ExitOpen;
                    }
                    fp->cluster = fp->ccls;
```

Next, we need to find an available entry space in the directory for the new file. This will require a second pass through the root directory, this time looking for the first entry that is either marked as deleted (code 0xE5) or for the end of the list where an empty entry is found (marked with the code 0x00).

```
// 11.2 create a new entry
// search again, for an empty entry this time
if ( (r = newDIR( fp)) == FAIL)              // report any error
{
    FError = FE_IDE_ERROR;
    goto ExitOpen;
}
```

The function newDIR() will take care of finding an available entry and, similarly to the findDIR() function used before, will return one of three possible codes:

- FAIL, indicating a major problem occurred (or the card was removed)

- NOT_FOUND, indicating the root directory must be full

- FOUND, indicating an available entry has been identified

```
// 11.3 new entry not found
if ( r == NOT_FOUND)
{
    FError = FE_DIR_FULL;
    goto ExitOpen;
}
```

In both the first two cases we have to report an error and we cannot continue. But if an entry is found, we have plenty of work to do to initialize it.

After calculating the offset of the entry in the current buffer, we will start filling some of its fields with data from the MFILE structure. The file size will be first:

```
else // 11.4 new entry identified fp->entry filled
{
    // 11.4.1 init file size
    fp->size = 0;

    // 11.4.2    determine offset in DIR sector
    e = (fp->entry & 0xf) * DIR_ESIZE;     // 16 entry per sector

    // 11.4.3 set initial file size to 0
    fp->buffer[ e + DIR_SIZE]   = 0;
    fp->buffer[ e + DIR_SIZE+1] = 0;
    fp->buffer[ e + DIR_SIZE+2] = 0;
    fp->buffer[ e + DIR_SIZE+3] = 0;
```

The time and date fields could be derived from the RTCC module registers or any other timekeeping mechanism available to the application, but a default value will be supplied here only for demonstration purposes.

```
    fp->date = 0x34FE; // July 30th, 2006
    fp->buffer[ e + DIR_DATE]   = fp->date;
```

```
fp->buffer[ e + DIR_DATE+1]= fp->date>>8;
fp->buffer[ e + DIR_TIME]  = fp->time;
fp->buffer[ e + DIR_TIME+1]= fp->time>>8;
```

The file's first cluster number, the file name and the attributes (defaults) will complete the directory entry:

```
// 11.4.5 set first cluster
fp->buffer[ e + DIR_CLST]  = fp->cluster;
fp->buffer[ e + DIR_CLST+1]= (fp->cluster>>8);

// 11.4.6 set name
for ( i = 0; i<DIR_ATTRIB; i++)
    fp->buffer[ e + i] = fp->name[i];

// 11.4.7 set attrib
fp->buffer[ e + DIR_ATTRIB] = ATT_ARC;

// 11.4.8  update the directory sector;
if ( !writeDIR( fp, fp->entry))
{
    FError = FE_IDE_ERROR;
    goto ExitOpen;
}
    } // new entry
} // not found
```

Back to the results of our first search through the root directory—in case a file with the same name was indeed found, we will need to report an error.

```
else // file exist already, report error
{
    FError = FE_FILE_OVERWRITE;
    goto ExitOpen
}
```

Alternatively, we would have had to delete the current entry first, release all the clusters used and then start from the beginning. After all, reporting the problem as an error is an easier way out for now.

So much for the changes required to the `fopenM()` function. We can now start writing the proper new `fwriteM()` function, once more modeled after a similarly named standard C library function.

```
unsigned fwriteM( void *src, unsigned count, MFILE * fp)
// src      points to source data (buffer)
// count    number of bytes to write
// returns  number of bytes actually written
{
    MEDIA *mda = fp->mda;
    unsigned len, size = count;
```

```
    // 1.  check if file is open
    if ( fp->entry + fp->name[0] != ~fp->chk )
    {   // checksum fails
        FError = FE_INVALID_FILE;
        return FAIL;
    }
```

The parameters passed to the function are identical to those used in the freadM() function and the first test we will perform on the integrity of the MFILE structure, passed as a parameter, is the same as well. It will help us determine if we can trust the contents of the MFILE structure, having been successfully prepared for us by a call to fopenM().

The core of the function will be a loop as well:

```
    // 2. loop writing count bytes
    while ( count>0)
    {
```

Our intention is to transfer as many bytes of data as possible at a time, using the fast memcpy() function from the "string.h" libraries.

```
        // 2.1 copy as many bytes at a time as possible
        if ( fp->pos+count < 512)
            len = count;
        else
            len = 512- fp->pos ;

        memcpy( fp->buffer+ fp->pos, src, len);
```

There are numerous pointers and counters that we need to update to keep track of our position as we add data to the buffer and we increase the size of the file.

```
        // 2.2 update all pointers and counters
        fp->pos+=len;       // advance buffer position
        fp->seek+=len;      // count the added bytes
        count-=len;         // update the counter
        src+=len;           // advance the source pointer

        // 2.3 update the file size too
        if (fp->seek > fp->size)
            fp->size = fp->seek;
```

Once the buffer is full, we need to transfer the data to the media in a sector of the currently allocated cluster.

```
        // 2.4 if buffer full, write current buffer to current sector
        if (fp->pos == 512)
        {
            // 2.4.1 write buffer full of data
            if ( !writeDATA( fp))
                return FAIL;
```

Notice that an error at this point would be rather fatal. We will return the code FAIL, whose value is 0, therefore indicating that not a single byte has been transferred; in fact all the data written to the storage device this far is now lost.

If all proceeds correctly, though, we can now increment the sector pointers and, if we have exhausted all the sectors in the current cluster, we need to consider the need to allocate a new one, calling new-FAT() once more.

```
            // 2.4.2 advance to next sector in cluster
            fp->pos = 0;
            fp->sec++;

            // 2.4.3 get a new cluster if necessary
            if ( fp->sec == mda->sxc)
            {
                fp->sec = 0;
                if ( newFAT( fp)== FAIL)
                    return FAIL;
            }
        } //  store sector

    } // while count
```

Shortly, when developing newFAT(), we will have to make sure that the function accurately maintains the chaining of the clusters in the FAT as they are added to a file.

```
    // 3.  number of bytes actually written

    return size-count;

} // fwriteM
```

The function is now complete and we can report the number of bytes written upon exit from the loop.

Closing a file, second take

While closing a file opened for reading was a mere formality and a matter of releasing some memory from the heap, when we close a file that has been opened for writing, there is a considerable amount of housekeeping work that needs to be performed.

A new and improved fcloseM() function is needed and it will start with a check of the mode field.

```
unsigned fcloseM( MFILE *fp)
{
    unsigned e, r;           // offset of directory entry in current buffer

    r = FAIL;

    // 1. check if it was open for write
    if ( fp->mode == 'w')
    {
```

In fact, when we close a file, there might still be some data in the buffer that needs to be written to the storage device, although it does not fill an entire sector.

```
// 1.1 if the current buffer contains data, flush it
if ( fp->pos >0)
{
    if ( !writeDATA( fp))
        goto ExitClose;
}
```

Once more, any error at this point is a rather fatal event and will mean that all the file data is lost since the `fcloseM()` function will not properly complete.

The proper root directory sector must be retrieved and an offset for the directory entry must be calculated inside the buffer.

```
// 1.2      finally update the dir entry,
// 1.2.1     retrive the dir sector
if ( !readDIR( fp, fp->entry))
    goto ExitClose;

// 1.2.2    determine position in DIR sector
e = (fp->entry & 0xf) * DIR_ESIZE;    // 16 entry per sector
```

Next, we need to update the file entry in the root directory with the actual file size (it had been initially set to zero).

```
// 1.2.3 update file size
fp->buffer[ e + DIR_SIZE]  = fp->size;
fp->buffer[ e + DIR_SIZE+1]= fp->size>>8;
fp->buffer[ e + DIR_SIZE+2]= fp->size>>16;
fp->buffer[ e + DIR_SIZE+3]= fp->size>>24;
```

Finally, the entire root directory sector containing the entry is written back to the media.

```
// 1.2.4     update the directory sector;
if ( !writeDIR( fp, fp->entry))
    goto ExitClose;
} // write
```

If all went well, we will complete the `fcloseM()` function invalidating the checksum field to prevent accidental reuses of this `MFILE` structure and deallocating the memory used by it and its buffer.

```
// 2. exit with success
r = TRUE;

ExitClose:
// 3. invalidate the file structure
fp->chk = fp->entry + fp->name[0];    // set checksum wrong!
```

```
    // 4. free up the buffer and the MFILE struct
    free( fp->buffer);
    free( fp);

    return( r);

} // fcloseM
```

Accessory functions

In completing fopenM(), fcloseM() and creating the new fwriteM() function, we have used a number of lower-level functions to perform important repetitive tasks.

We will start with newDIR(), used to find an available spot in the root directory to create a new file. The similarity with findDIR() is obvious, yet the task performed is very different.

```
unsigned newDIR( MFILE *fp)
// fp         file structure
// return     found/fail,   fp->entry filled
{
    unsigned eCount;            // current entry counter
    unsigned e;                 // current entry offset in buffer
    int a;
    MEDIA *mda = fp->mda;

    // 1. start from the first entry
    eCount = 0;
    // load the first sector of root
    if ( !readDIR( fp, eCount))
        return FAIL;

    // 2. loop until you reach the end or find the file
    while ( 1)
    {
    // 2.0 determine the offset in current buffer
        e = (eCount&0xf) * DIR_ESIZE;

    // 2.1 read the first char of the file name
        a = fp->buffer[ e + DIR_NAME];

    // 2.2 terminate if it is empty (end of the list) or deleted
        if (( a == DIR_EMPTY) ||( a == DIR_DEL))
        {
            fp->entry = eCount;
            return FOUND;
        } // empty or deleted entry found
```

```
        // 2.3 get the next entry
            eCount++;
            if ( (eCount & 0xf) == 0)
            { // load a new sector from the root
                if ( !readDIR( fp, eCount))
                    return FAIL;
            }

        // 2.4 exit the loop if reached the end or error
            if ( eCount > mda->maxroot)
                return NOT_FOUND;                 // last entry reached
        }// while

        return FAIL;
    } // newDIR
```

The function newFAT() was used to find an available cluster to allocate for a new block of data/new file.

```
unsigned newFAT( MFILE * fp)
// fp          file structure
// fp->ccls    ==0 if first cluster to be allocated
//             !=0 if additional cluster
// return      TRUE/FAIL
//  fp->ccls new cluster number
{
    unsigned i, c = fp->ccls;

    // sequentially scan through the FAT looking for an empty cluster
    do {
        c++;      // check next cluster in FAT
        // check if reached last cluster in FAT, re-start from top
        if ( c >= fp->mda->maxcls)
            c = 0;

        // check if full circle done, media full
        if ( c == fp->ccls)
        {
            FError = FE_MEDIA_FULL;
            return FAIL;
        }

        // look at its value
        i = readFAT( fp, c);
```

```
    } while ( i!=0);     // scanning for an empty cluster



    // mark the cluster as taken, and last in chain
    writeFAT( fp, c, FAT_EOF);

    // if not first cluster, link current cluster to the new one
    if ( fp->ccls >0)
        writeFAT( fp, fp->ccls, c);

    // update the MFILE structure
    fp->ccls = c;

    return TRUE;
} // allocate new cluster
```

When allocating a new cluster beyond the first one, newFAT() keeps linking the clusters in a chain and it marks every cluster as properly used. For its working, the function uses two accessory functions readFAT() and writeFAT() in turn.

```
unsigned readFAT( MFILE *fp, unsigned ccls)
// fp        MFILE structure
// ccls      current cluster
// return    next cluster value,
//           0xffff if failed or last
{
    unsigned p, c;
    LBA l;

    // get address of current cluster in fat
    p = ccls;
    // cluster = 0xabcd
    // packed as:     0  |  1    |  2   |  3    |
    // word p       0  1 | 2   3 | 4   5 | 6   7 |..
    //              cd ab| cd ab| cd ab| cd ab|

    // load the fat sector containing the cluster
    l = fp->mda->fat + (p >> 8 );   // 256 clusters per sector
    if ( !readSECTOR( l, fp->buffer))
        return 0xffff;  // failed

    // get the next cluster value
    c = ReadOddW( fp->buffer, ((p & 0xFF)<<1));

    return c;

} // readFAT
```

The `writeFAT()` function updates the contents of the FAT and keeps all its copies current.

```
unsigned writeFAT( MFILE *fp, unsigned cls, unsigned v)
// fp        MFILE structure
// cls       current cluster
// v         next value
// return    TRUE if successful, or FAIL
{
    unsigned p;
    LBA l;

    // get address of current cluster in fat
    p = cls * 2; // always even
    // cluster = 0xabcd
    // packed as:      0  |  1   |   2  |  3   |
    // word p          0  1 |  2   3 |  4   5 |  6   7 |..
    //                 cd ab| cd ab| cd ab| cd ab|

    // load the fat sector containing the cluster
    l = fp->mda->fat + (p >> 9 );
    p &= 0x1fe;
    if ( !readSECTOR( l, fp->buffer))
        return FAIL;

    // get the next cluster value
    fp->buffer[ p] = v;         // lsb
    fp->buffer[ p+1] = (v>>8);// msb

    // update all FAT copies
    for ( i=0; i<fp->mda->fatcopy; i++, l += fp->mda->fatsize)
        if ( !writeSECTOR( l, fp->buffer))
            return FAIL;

    return TRUE;

} // writeFAT
```

Finally, `writeDATA()` was used both by `fwriteM()` and `fcloseM()` to write actual sectors of data to the storage device, computing the sector address based on the current cluster number.

```
unsigned writeDATA( MFILE *fp)
{
    LBA l;

    // calculate lba of cluster/sector
    l = fp->mda->data + (LBA)(fp->ccls-2) * fp->mda->sxc + fp->sec;

    return ( writeSECTOR( l, fp->buffer));

} // writeDATA
```

Testing the complete fileio module

It is time to test the functionality of the entire module we just completed. As in the previous test, we will use the serial communication module "`conu2.c`" developed in one of the early lessons and the same delay module that will provide a `delayms()` function. This time, after mounting the file system, we will open a source file (that could be any file), and we will copy its contents into a new "destination" file that we will create on the spot. Here is the code we will use for the "`writetest.c`" main file.

```c
/*
** WriteTest.c
**
*/

#include <p24fj128ga010.h>

#include "SDMMC.h"
#include "fileio.h"
#include "../delay/delay.h"

#include "../8 comm/conu2.h"

#define B_SIZE   1024

char data[B_SIZE];


int main( void)
{
    MFILE *fs, *fd;
    unsigned r;

    //initializations
    initU2();            //115,200 baud 8,n,1

    putsU2( "init");
    while( !detectSD());   // assumes SDCD pin is by default an input
    Delayms( 100);         // wait for card to power up

    if ( mount())
    {
        putsU2("mount");
        if ( (fs = fopenM( "source.txt", "r")))
        {
            putsU2("source file opened for reading");
            if ( (fd = fopenM( "dest3.txt", "w")))
            {
```

```
                    putsU2("destination file opened for writing");
                    do{
                        r = freadM( data, B_SIZE, fs);

                        r = fwriteM( data, r, fd);

                        putU2('.');

                        } while( r==B_SIZE);

                        fcloseM( fd);
                        putsU2("destination file closed");
                    }
                    else
                        putsU2("could not open the destination file");

                    fcloseM( fs);
                    putsU2("source file closed");
                }
                else
                    putsU2("could not open the source file");

                unmount();
                putsU2("unmount");

            }
            else
                putsU2("mount failed");

            // main loop
            while( 1);
        } // main
```

Make sure you replace the source file name with the actual name of the file you copied on the card for the experiment.

After creating a new project (let's call it "WriteTest" this time), we will need to add all the necessary modules to the project window, including:

- "sdmmc.c"

- "fileio.c"

- "conu2.c"

- "delay.c"

- "writetest.c"

and all the corresponding include files (.h).

Once more, remember to follow the checklists for a new project and for the ICD2 debugger, but this time remember to add some more space for the heap so that we will be able to allocate dynamically at least two buffers and two `MFILE` structures.

> Note: Once enough space is left for the global variables and the stack, there is no reason to withhold any memory from the heap. Allocate as large a heap as possible to allow `malloc()` and `free()` to make optimal use of all the memory available.

After building the project and programming the executable on the Explorer16 board, we are ready to run the test. If all goes well, after a fraction of a second (the actual time will depend on the size of the source file chosen) you will be able to see on the screen of your terminal the following messages:

```
init
mount
source file opened for reading
destination file opened for writing
....................................................................
destination file closed
source file closed
unmount
```

The number of dots will be proportional to the size of the file, and since we chose the buffer size to be 1024 for this demo, each dot will correspond exactly to one kilobyte of data transferred. At this point, if you transfer the SD/MMC card back to your PC, you should be able to verify that a new file has been created.

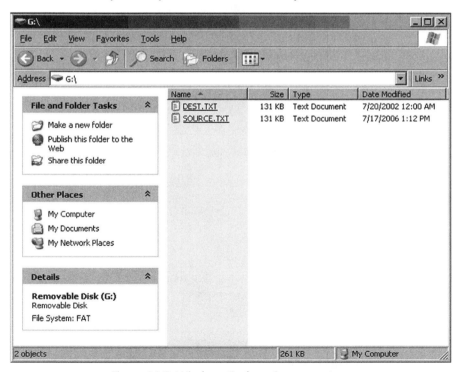

Figure 14-8. Windows Explorer Screen capture.

Its size and content are identical to the source file, while the date and time reflect the values we set in the `fcloseM()` function.

Notice that if you try to run the test program a second time, it is bound to fail now.

```
init
mount
source file opened for reading
could not open the destination file
source file closed
unmount
```

This is because, as discussed during the development of the `fopenM()` function, we chose to report an error when trying to open a file for writing and on the storage device finding a file (`DEST.TXT`) already there with the same name.

Notice that you can recompile the project and run the test with different sizes for the data buffer, from 1 byte to as large as the memory of the PIC24 will allow. Both the `freadM()` and `fwriteM()` functions will take care of reading and writing as many sectors of data as required to fulfill your request. The time required to complete the operation will change though.

Code Size

The size of the code produced by the "WriteTest" project is considerably larger than the simple "`sdmmc.c`" module we tested in the previous lesson.

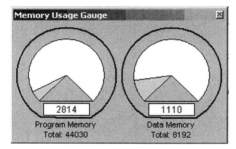

Figure 14-9. The Memory Usage Gauge.

Still, with all optimization options turned off, the code will add up to just 8,442 bytes (2814 words * 3). This represents only 6% of the total program memory space available on the PIC24FJ128GA010. I consider this a very small price to pay for a lot of functionality!

Post-flight briefing

In this lesson we have learned the basics of the FAT16 file system and we have developed a small interface module that allows a PIC24 to read and write data files to and from a generic mass-storage device. By using the "`sdmmc.c`" module, developed in the previous lesson for the low-level interface, we have created a basic file I/O interface for SD/MMC memory cards.

Now you can share data between the PIC24 and most any other computer system that is capable of accessing SD/MMC cards, from PDAs to laptops and desktop PCs, from DOS, Windows and Linux machines to Apple computers running OS-X.

Tips and tricks

A frequent question I get from embedded-control engineers is: "How can I interface to a "Thumb drive" (sometimes referred to as a USB stick), a USB mass-storage device, to share/transport data between my application and a PC?"

The short answer I have is simple: "Don't, if you can help it!"

The longer answer is: "Use an SD card instead!" and here is why. As you have seen in this lesson and the previous one, reading and writing to an SD card is really simple and requires very little code and only an SPI port (possibly shared, too).

The USB interface, on the other hand, has all the appeal and appearance of simplicity from the user perspective, but reading and writing to a USB thumb drive can be deceptively complex and expensive for a modest embedded-control application. First of all, the simplicity of the SPI interface must be replaced by the relatively greater complexity of the USB bus interface. What is required, then, is not just the standard USB peripheral kind of interface, but the Host USB protocol in its full glory. There is a considerable hardware price to pay, in terms of dedicated USB transceivers and a large Serial Interface Engine (SIE) module required. There is an even larger cost in terms of the code and RAM memory required to support it all. This can be estimated to be several orders of magnitude larger than the basic SD/MMC card solution we have examined here.

Exercises

1. Consider adding the following functionality:
   – Subdirectories management.
   – Erasing files.
   – Long file-name support.

2. Use the RTCC to provide the current time and date information

3. Consider caching (and/or using a separate buffer) for the current FAT record content to improve read/write performance

4. Consider the modifications required to perform buffering of larger blocks and/or entire clusters and performing multiblock read/write operations to optimize the SD card low-level performance. Consider pros and cons.

Books

- Buck, B. (2002)

  North Star Over My Shoulder

  Simon & Shuster, New York, NY

  The story of aviation through the experiences of a lifetime as a pilot.

Links

- *http://www.tldp.org/LDP/tlk/tlk-title.html*

  "The Linux Kernel" by David A. Rusling, an online book that describes the inner workings of Linux and its file system.

- *http://en.wikipedia.org/wiki/File_Allocation_Table*

  Once more, an excellent page of the wikipedia that describes the history and many ramifications of the FAT technology.

- *http://en.wikipedia.org/wiki/List_of_file_systems*

  An attempt to list and classify all major computer file systems in use.

Volare

The last flight, the check-ride with the FAA examiner, is a time of great tension and a little fear. It is a flight meant to summarize all the phases of flight, where you are asked to put all the knowledge you gained during the training into practice. Don't worry—it will be easy because you are at the peak of your preparation and it will be over so fast that you won't have time to realize it.

Just as in a final check-ride, this last lesson will use many of the building blocks developed in the previous lessons and will put them to practical use to develop a new and entertaining demo project: a media player.

Congratulations, you are a pilot now. It is time to celebrate and sing!

NEL BLU DIPINTO DI BLU
Italy 1958 / Domenico Modugno
Written by Franco Migliacci & Domenico Modugno

Penso che un sogno cosí non ritorni mai piú:
Mi dipingevo le mani e la faccia di blu
Poi d'improvviso venivo dal vento rapito
E incominciavo a volare nel cielo infinito
Volare, oh...cantare, oh...

The lyrics are in Italian. The title translates to "In The Blue (Sky), Painted in Blue" ("Volare": to fly). Modugno cowrote it with Franco Migliacci after Modugno described a man's dream of flying through the air with his face and hands painted in blue.

Flight plan

In this lesson we will explore the possibility of producing audio signals using, once more, the Output Compare modules of the PIC24. When in the Pulse Width Modulation (PWM) mode in combination with more or less sophisticated low-pass filters, the Output Compare modules can be used effectively as digital-to-analog converters to produce an analog output signal. If we manage to modulate the analog signal with frequencies that fall into the range that is recognized by the human ear, approximately between 20 Hz and 20 kHz, we get sound!

The flight

The way a pulse width modulation signal works is pretty simple. A pulse is produced at regular intervals (T) typically provided by a timer and its period register. The pulse width (T_{on}) though is not fixed, but it is programmable and it can vary between 0 and 100% of the timer period. The ratio between the pulse width (T_{on}) and the signal period (T) is called the duty cycle.

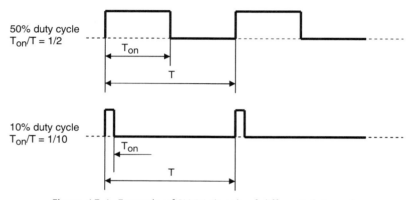

Figure 15-1. Example of PWM signals of different duty cycle.

There are two extreme cases possible for the duty cycle: 0% and 100%. The first one corresponds to a signal that is always off. The second one is the case when the output signal is always on. The number of possible cases in between, typically a relatively small finite number expressed as a logarithm in base 2, is commonly referred to as the resolution of the PWM. If, for example, there are 256 possible pulse widths, we will say that we have a PWM signal with an 8-bit resolution.

If you could feed an ideal PWM signal with fixed duty cycle to a spectrum analyzer to study its composition, you would discover that it contains three parts:

- a DC component, with an amplitude directly proportional to the duty cycle.

- a sinusoid at the fundamental frequency ($f = 1/T$).

- followed by an infinite number of harmonics whose frequency is a multiple of the fundamental ($2f$, $3f$, $4f$, $5f$, $6f$...).

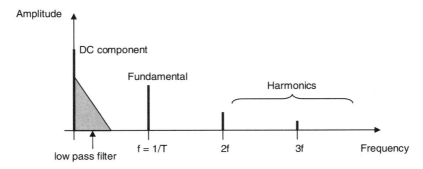

Figure 15-2. Frequency spectrum of a PWM signal.

Therefore, if we could attach an "ideal" low-pass filter to the output of a PWM signal generator to remove all frequencies from the fundamental and up, we could obtain a clean DC analog signal whose amplitude would be directly proportional to the duty cycle.

Of course such an ideal filter does not exist, but we can use more or less sophisticated approximations of it to remove as much of the unwanted frequency components as needed. This filter could be as simple as a single passive R/C circuit (first-order low-pass filter) or could require several (N) active stages ($2 \times N$-order low pass).

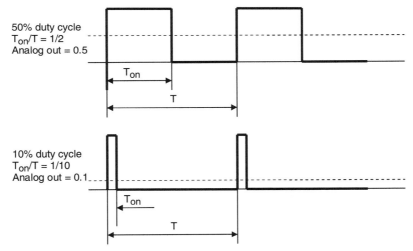

Figure 15-3. Analog output of PWM and ideal low pass filter circuit.

If we aim at producing an audio signal and we choose the PWM frequency wisely, we can take advantage of the natural limitation of the human ear that will act as an additional filter ignoring any signal whose frequency is outside the 20 Hz to 20 kHz range. In addition to that, most of the audio amplifiers we might want to feed the output signal into will also include a similar type of filter in their input stages. In other words, if we make sure that the PWM signal operates on a frequency at or above 20 kHz, both phenomenon will contribute to help our cause and will allow us to use a simpler and more inexpensive filter circuit.

Also intuitively enough, since we can only change the duty cycle once every PWM period (T), the higher the frequency of the PWM, the faster we will be able to change the output analog signal, and therefore the higher will be the frequency of the audio signal we will be able to generate.

In practical terms, this means that the highest audio signal a PWM can produce is only half of the PWM frequency. So, for example, a 20-kHz PWM circuit will be able to reproduce only audio signals up to 10 kHz, while to cover the entire audible frequency spectrum we need a base period of at least 40 kHz. It is not a coincidence that, for example, music CDs are digitally encoded at the rate of 44,100 samples per second.

Using the PIC24 OC modules in PWM mode

In a previous lesson we have already used the PIC24 Output Compare modules to produce precise timing intervals (to produce a video output). This time we will use the OC modules in PWM mode, to generate a continuous stream of pulses with the desired duty cycle.

Upper Byte:

U-0	U-0	R/W-0	U-0	U-0	U-0	U-0	U-0
—	—	ICSIDL	—	—	—	—	—
bit 15							bit 8

Lower Byte:

R/W-0	R/W-0	R/W-0	R-0, HC	R-0, HC	R/W-0	R/W-0	R/W-0
ICTMR	ICI1	ICI0	ICOV	ICBNE	ICM2	ICM1	ICM0
bit 7							bit 0

Figure 15-4. The Output Compare module main control register OCxCON.

All we need to do to initialize the OC module to generate a PWM signal is set the three OCM bits in the OCxCON control register to the 0x110 configuration. A second PWM mode is available (0x111), but we have no use for the fault input pins (OCFA/OCFB), commonly required by a different set of applications as a protection mechanism (motor control/power conversion). Next we need to select the timer on which to base the PWM period. The choice is limited to Timer2 or Timer3, and for now it will make no difference to us. It is how we will configure the chosen timer that will make all the difference. (See Figure 15-5.)

Keeping in mind that we want to be able to produce at least a 40 kHz PWM period, and assuming a peripheral clock of 16 MHz as is the case when using the Explorer16 board, we can calculate the optimal values for the prescaler and the period register PRx. With a prescaler set to a 1:1 ratio, we obtain a 400-cycle period, generating an exact 40-kHz signal. This value also dictates the resolution of the duty cycle for the Output Compare module. Since we will have 400 possible values of the duty cycle, we can claim a resolution between 8 and 9 bits, as we have more than 256 (2^8) steps but less than 512 (2^9). Reducing the frequency to 20 kHz would give us one bit more of resolution (between 9 and 10), but would also mean that we would be limiting the output frequency range to a maximum of 10 kHz, probably a small but noticeable difference to the human ear. Once the chosen timer is configured and just before writing to the OCxCON register, we will need to set, for the first time, the value of the first duty cycle writing to the register OCxR, and the register OCxRS. When in PWM mode, the two registers will work in a master/slave configuration. Once the PWM module is started (writing the mode bits in the OCxCON register), we will be able to change the duty cycle by writing only to the OCxRS register. The OCxR register will be updated, copying a new value from OCxRS, only and precisely at the beginning of each new period so as to avoid glitches and to leave us with an entire period (*T*) of time to prepare the next duty cycle value:

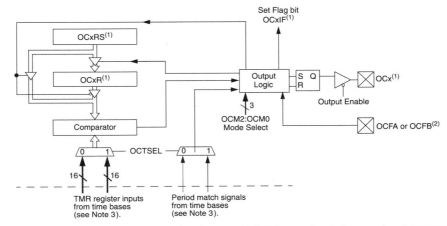

Figure 15-5. Output Compare module block diagram.

Here is an example of a simple initialization routine for the OC1 module:

```
void initAudio( void)
{
    // init TMR3 to provide the timebase
    T3CON = 0x8000;        // enable TMR3, prescale 1:1, internal clock
    PR3 = 400-1;           // set the period for the given bitrate
    _T3IF = 0;             // clear interrupt flag
    _T3IE = 1;             // enable TMR3 interrupt

    // init PWM
    // set the initial duty cycles (master and slave)
    OC1R = OC1RS = 200;         // init at 50%

    // activate the PWM module
    OC1CON = 0x000E;

} // initAudio
```

Notice that we have also taken the opportunity to enable the Timer3 interrupt so that we are alerted each time a new period starts, and we can decide how and if to update the next duty cycle value writing to OC1RS.

Testing the PWM as a D/A converter

To start experimenting on the Explorer16 we will need to add just a couple of discrete components to the prototyping area. A 1-kohm resistor and a 100 nF capacitor will produce the simplest low-pass filter (first order with a 1.5-kHz cut-off frequency). We can connect the two in series and wire them to the output pin of the OC1 module found on pin 0 of PORTD as represented in the schematic below.

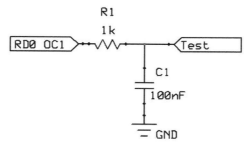

Figure 15-6. Using a PWM signal to produce an analog output.

A couple of more lines of code will complete our short test project:

```
void _ISRFAST _T3Interrupt( void)
{
    // clear interrupt flag and exit
    _T3IF = 0;
} // T3 Interrupt


main( void)
{
    initAudio();

    // main loop
    while( 1);

}// main
```

Add the usual header and include file, and save the code in a new file called TestDA.c. You can then create a quick test project that will contain this single file, build it and use the ICD2 debugger to program the Explorer16 board.

After connecting a meter, or an oscilloscope probe if available, to the test point, run the program and verify the average (DC) output level.

The needle of the meter (or the trace of the scope) will swing to a voltage level of approximately 1.5V that is 50% of the regular voltage output of a digital I/O pin on the Explorer16 board. This is consistent with the value of the duty cycle set by the initialization routine to 200 (for a period of 400 cycles). If you have an oscilloscope, you can also point the probe directly at the other end of the R1 resistor (directly to the output pin of the OC1 module) and verify that a square wave of the exact frequency of 40 kHz is present with a duty cycle of 50%.

You can now change the initialization routine to experiment with other values between 0 and 399 to verify the response of the circuit and the proportionality of the output signal to the changing values of the duty cycle with analog values between 0 and 3V.

Producing analog waveforms

With help from the OC1 module, we have just crossed the boundary between the digital world made of ones and zeros and the analog world in which we have been able to generate a multitude of values between 0 and 3V.

We can now play with the duty cycle, changing it from period to period to produce waveforms of any sort and shape. Let's start by modifying the project a little bit, adding some code to the interrupt routine that so far has been left empty:

```
OC1RS = (count < 20) 400 : 0;
count++;
if ( count >= 40)
    count = 0;
```

You will need to declare count as a global integer and remember to initialize it to 0.

Save and rebuild the project to test the new code on the Explorer16 board.

Every 20 PWM periods, the filter output will alternate between the value 3V (100%) and the value 0V (0%) producing a square wave visible on the oscilloscope at the frequency of 1 KHz (40 kHz/40).

A more interesting waveform could be generated by the following algorithm:

```
OC1RS = count*10;
count++;
if ( count >= 40)
    count = 0;
```

This will produce a triangular waveform (sawtooth) of approximately 3V peak amplitude, with a gradual ramp of the duty cycle from 0 to 100% in 40 steps (2.5% each), followed by an abrupt fall back to 0 where it will repeat. This signal will repeat with a frequency of 1 kHz as well.

Neither of the two examples will qualify as a "nice" sound if you try to feed them to an audio amplifier, although they will both have a recognizable (fundamental) high-pitched tone, at about 1 kHz. Lots of harmonics will be present and audible in the audio spectrum and will give the sound an unpleasant buzz.

To generate a single clean tone, what we need is a pure sinusoid. The interrupt service routine below would serve the purpose, generating a perfect sinusoid at the frequency of 400 Hz (in musical terms, that would be close to an A).

```
void _ISRFAST _T3Interrupt( void)
{// compute the new sample for the next cycle
    OC1RS = 200+ ( 200* sin(count *0.0628));
    count++;
    if ( count >= 40)
        count = 0;
    // clear interrupt flag and exit
    _T3IF = 0;
} // T3 Interrupt
```

Unfortunately, as fast as the PIC24 and the math libraries of the C30 compiler are, there is no chance that we can use the sin() function, perform the multiplications and additions required, and obtain a new duty-cycle value at the required rate of 400 Hz. The Timer3 interrupt hits every 25 µs, too short a time for such a complex floating-point calculation, so the interrupt service routine would end up "skipping" interrupts and producing a sinusoidal output that is only half the required frequency (one octave lower). Still, if you listen to it by feeding the signal to an audio amplifier, you will be immediately able to appreciate the greatly improved clarity of the sound.

For higher frequencies we will need to pretabulate the sinusoid values so as to perform the smallest number of calculations possible (preferably working on integers only) at run time. Here is an example that uses a table (stored in the Flash program memory of the PIC24) containing precomputed values. I obtained the table by using a spreadsheet program in which I used the following formula:

```
= offset + INT( amplitude * SIN( ROW * 6.28/ PERIOD))
```

For a period of 100 samples (400 Hz), offset and amplitude of 200, I obtained:

```
=200 + INT( 200*SIN(A1 *6.28/100))
```

I filled the first column (A) of the spreadsheet with a counter and I copied the formula over the first 100 rows of the second column (B), formatting the output for zero decimal digits.

Then, I cut and pasted the entire column in the source code, adding commas at the end of each line to comply with the C syntax.

```c
void _ISRFAST _T3Interrupt( void)
{
    // load the new samples for the next cycle
    OC1RS = Table[ count];

    count++;
    if ( count >= 40)
        count = 0;

    // clear interrupt flag and exit
    _T3IF = 0;
} // T3 Interrupt


const int Table[100] = {
200,
212,
225,
237,
249,
...
149,
161,
174,
186,
199
};
```

This time, you will easily be able to produce the 400-Hz output frequency desired, and there will be plenty of time between the Timer3 interrupt calls to perform other tasks as well.

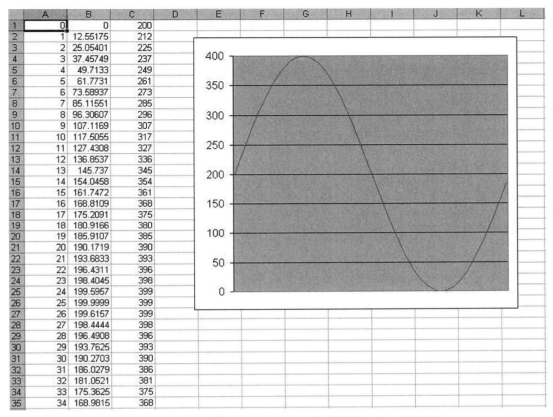

Figure 15-7. Spreedsheet to compute the 400-Hz sinusoid table.

Reproducing voice messages

Once we learn how to produce sound, there is no stopping us. There are infinite applications in embedded control where we can put these capabilities to use. Any "human" interface can be greatly enhanced by using sound to provide feedback, to capture the attention of the user with alerts and error messages or, if properly done, to simply enhance the user experience. But we don't have to limit ourselves to simple tones or basic melodies. We can, in fact, reproduce any kind of sound, as long as we have a description of the waveforms required. Just like the table used for the sinusoid in the previous example, we could use a larger table to contain the unmistakable sound produced by a particular instrument or even a complete vocal message. The only limit becomes the room available in the Flash program memory of the PIC24 to store the data tables next to the application code.

If, in particular, we look at the possibility of storing voice messages, knowing that the energy of the human voice is mostly concentrated in the frequency range between 400 Hz and 4 kHz, we can considerably reduce our output frequency requirements and limit the PWM playback to the rate of only 8,000 samples per second. Notice that we should still maintain a high PWM frequency to keep the PWM signal harmonics outside the audio frequency range and the low-pass filter simple and inexpensive. It is only the rate at which we change the PWM duty cycle and read new data from the table that will have to be reduced, once every five interrupts in this case (40,000/8,000 = 5). With 8,000 samples per second, we would theoretically be able to play back as much as 16 seconds of voice messages stored inside the controller Flash memory. That is already a lot of talking for a single-chip solution. To increase the capacity further, potentially doubling it, we could start looking at simple compression techniques used for voice applications such as ADPCM, for example. ADPCM stands for Adaptive Differential Pulse Coded Modulation, and it is based on the assumption that the difference between two consecutive samples is smaller than the absolute value of each sample and can therefore be encoded using a smaller number of bits. The actual number of bits used is then optimized; it changes dynamically so as to avoid signal distortion while providing a desired compression ratio—hence the use of the term "adaptive."

A media player

In the rest of this lesson, we will focus on a much more ambitious project, putting to use all the libraries and capabilities we have acquired in the last several lessons. We will attempt to create a basic multimedia application capable of playing stereo music files from an SD™/MMC memory card. The format of choice will be the uncompressed WAVE format that is compatible with almost any audio application and is the default destination format when extracting files from a music CD.

We will start by creating a brand new project using the usual checklists. We will immediately add the SD/MMC low-level interface and the file I/O library for access to a FAT16 file system to the project source files list.

After opening a file for reading, though, this time we will need to be able to understand the specific format used to encode the data it contains.

The WAVE file format

Files with the .wav extension, encoded in the WAVE format, are among the simplest and best documented, but they still require some careful study. The WAVE format is a variant of the RIFF file format, a standard across multiple operating systems, which uses a particular technique to store multiple pieces of information/data by dividing them into "chunks." A chunk is nothing more than a block of data preceded by a header containing two 32-bit elements: the chunk ID and the chunk size.

Offset	Size	Value	Description
0x00	4	ASCII	Chunk ID
0x04	4	Size	Chunk size (size of the content)
0x08	size		Data content
0x08+size	1	0x00	Optional padding

Table 15-1. Format of a data "chunk."

Note also that the chunk total size must be a multiple of two so that all the data in a RIFF file ends up being nicely word aligned. If the data block size is not a multiple of two, an extra byte of padding is added to the chunk.

A chunk with the "RIFF" ID is always found at the beginning of a .wav file and its data block begins with a 4-byte "type" field. This type field must contain the string "WAVE". Chunks can be nested like Russian dolls, but there can also be multiple subchunks inside a given type of chunk.

The Table 15-2 illustrates a ".wav" file RIFF chunk structure:

Offset	Size	Value	Description
0x00	4	"RIFF"	This is the RIFF chunk ID
0x04	4	Size	(size of the data block+4)
0x08	4	"WAVE"	Type ID
0x10	Size		Data block (subchunks)

Table 15-2. "RIFF" chunk of type "WAVE".

The data block in its turn contains a "fmt" chunk followed by a "data" chunk. As is often the case, one image is worth a thousands words.

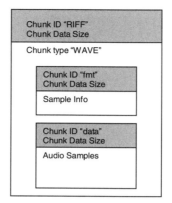

Figure 15-8. Basic WAVE file layout.

The "fmt" chunk contains a defined sequence of parameters that fully describes the stream of samples that follows in the "data" chunk, as represented in Table 15.3.

Offset	Size	Description	Value
0x00	4	Chunk ID	"fmt "
0x04	4	Chunk Data Size	16 + extra format bytes
0x08	2	Compression code	Unsigned int
0x0a	2	Number of channels	Unsigned int
0x0c	4	Sample rate	Unsigned long
0x10	4	Average bytes per second	Unsigned long
0x14	2	Block align	Unsigned int
0x16	2	Significant bits per sample	Unsigned int (>1)
0x18	2	Extra format bytes	Unsigned int

Table 15-3. The "fmt" chunk content.

In between the "fmt" and "data" chunks there could be other chunks containing additional information about the file, so we might have to scan the chunk IDs and skip through the list until we find the chunk we are looking for.

The `play()` function

Let's create a new software module that will take care of opening a given ".WAV" file and, after capturing and decoding the information in the "fmt" chunk, will initialize an audio output module similar to, if not more sophisticated than, the one we developed in the first part of the lesson. We will call it "wave.c".

```
/*-------------------------------------------------------------------
** Wave.C
**
**  Wave File Player
**  Uses 2 x 8 bit PWM channels
**
*/
#include <stdlib.h>
#include "../Audio/Audio PWM.h"
#include "../sdmmc/sdmmc.h"
#include "../sdmmc/fileio.h"

// chunk ID definitions
#define   RIFF_DWORD   0x46464952UL
#define   WAVE_DWORD   0x45564157UL
#define   DATA_DWORD   0x61746164UL
#define   FMT_DWORD    0x20746d66UL


typedef struct {
    // data chunk
    unsigned long dlength;  // actual data size
    char     data[4];       // "data"

    // format chunk
    unsigned     bitpsample; //
    unsigned     bpsample;   // bytes per sample (4 = 16 bit stereo)
    unsigned long bps;       // bytes per second
    unsigned long srate;     // sample rate in Hz
    unsigned     channels;   // number of channels (1= mono, 2= stereo)
    unsigned     subtype;    // always 01
    unsigned long flength;   // size of encapsulated block (16)
    char     fmt_[4];        // "fmt_"

    char     type[4];        // file type name "WAVE"
    unsigned long tlength;   // size of encapsulated block
    char     riff[4];        // envelope "RIFF"
} WAVE;
```

The WAVE structure will be useful to collect all the "fmt" parameters in one place and the chunk ID macros will help us recognize the different unique IDs, treating them as long integers and allowing us a quick and efficient comparison.

Next, let's start coding the `play()` function. It needs just one parameter: the file name.

```
unsigned play( const char *name)
{
    int     i;
    WAVE    wav;
    MFILE   *f;
    unsigned wi;
    unsigned long lc, r, d;
    int     skip, size, stereo, fix, pos;

    // 1. open the file
    if ( (f = fopenM( name, "r")) == NULL)
    {   // failed to open
        return FALSE;
    }
```

After trying to open the file and reporting the error if unable, we will immediately start looking inside the data buffer for the RIFF chunk ID and the WAVE type ID, as a signature, that will confirm to us we have the right kind of file:

```
// 2. verify it is a RIFF formatted file
if ( ReadL( f->buffer, 0) != RIFF_DWORD)
{
    fclose( f);
    return FALSE;
}

// 3. look for the WAVE type
if ( (d = ReadL( f->buffer, 8)) != WAVE_DWORD)
{
    fclose( f);
    return FALSE;
}
```

If successful, we should verify that the "fmt" chunk is the first in line inside the data block. Then we will harvest all the information needed to process the data block for the playback:

```
// 4. look for the chunk containing the wave format data
if ( ReadL( f->buffer, 12) != FMT_DWORD)
    return FALSE;

wav.channels   = ReadW( f->buffer, 22);
wav.bitpsample = ReadW( f->buffer, 34);
wav.srate      = ReadL( f->buffer, 24);
wav.bps        = ReadL( f->buffer, 28);
wav.bpsample   = ReadW( f->buffer, 32);
```

Next, we start looking for the "data" chunk, inspecting the chunk ID fields of the next block of data after the end of the "fmt" chunk, and skipping the entire block if not matching:

```
// 5. search for the data chunk
wi = 20 + ReadW( f->buffer, 16);
while ( wi < 512)
{
    if (ReadL( f->buffer, wi) == DATA_DWORD)
        break;
    wi += 8 + ReadW( f->buffer, wi+4);
}
if ( wi >= 512) // could not find a data chunk in current sector
{
    fclose( f);
    return FALSE;
}
```

If, in the process, we exhaust the content of the currently loaded buffer of data, we know we have a problem. Typical WAV files produced by extracting data from a music CD will have just the "data" chunk immediately following the "fmt" chunk. Other applications (MIDI interfaces, for example) can generate "WAV" files with more complex structures including multiple "data" chunks, "playlists." "cues." "labels", etc. but we aim at playing back only plain-vanilla type "WAV" files.

Once found, the size of the "data" chunk will tell us the real number of samples contained in the file:

```
// 5.1 find the data size (actual wave content)
wav.dlength = ReadL( f->buffer, wi+4);
```

The playback sample rate must now be taken into consideration to determine if we can "play" that fast. It could in fact exceed our capabilities, and we might have to skip every other sample to reduce the data rate. We will consider 48k samples/sec our limit so we will be able to read the data fast enough to maintain at least an 8-bit resolution.

```
// 6. compute the period and adjust the bit rate
r = wav.bps / wav.bpsample;    // r = samples per second
skip = wav.bpsample;           // skip factor to reduce bandwith (stereo)
while ( r > 48000)
{
    r >>= 1;                   // as you divide sample rate by two
    skip <<= 1;                // multiply skip by two
}
```

Higher rates will be treated by gradually dividing the rate by a factor of two and doubling the skip.

We can then compute the required PWM period value (to be used to set the PRx register). A problem could occur if the required period exceeds the available bits in the register (16), resulting in a period value greater than 65,536.

```
// 6.1 check if the sample rate is compatible with the TMR3 prescaler 1:1
d = (16000000L/r)-1;
if ( d > ( 65536L))                    // max TMR3 period value (16 bit)
{
    fclose( f);
    return FALSE;
}
```

During the playback we will keep track of the number of samples extracted from the file to determine when we have reached the end of the file. The long integer variable (`lc`) will keep track of it.

```
// 7. start loading the buffers
// determine the number of bytes composing the wav data chunk
lc = wav.dlength;
```

Notice that so far we have not used the `freadM()` function; we have been (cheating) peeking inside the file buffer knowing that `fopenM()` had it already loaded.

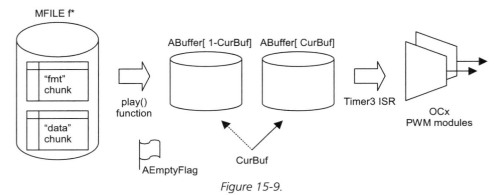

Figure 15-9.

To make the playback smooth, we will use a double-buffering scheme, so that as the audio interrupt routines are fetching data from one buffer, we will take our time in refilling the other buffer with new data from the file. The array `ABuffer[]` is in fact defined as two blocks of `B_SIZE` bytes each. `B_SIZE` is chosen to be a multiple of 512, so that the calls to the `freadM()` function will be able to transfer entire sectors of data at a time for maximum efficiency. We will have to verify that the time required for `freadM()` to fill one buffer will be shorter than the time required to play back (consume) all the data in the second buffer by the PWM interrupt service function.

When starting the double-buffering scheme, we will fill both buffers to get a head start:

```
// 8. pre-load both buffers
r = fread( ABuffer[0], B_SIZE*2, f);
AEmptyFlag = FALSE;
lc-= B_SIZE*2 ;                        // we assume here that lc>=B_SIZE*2!!!
```

The assumption here is that the .WAV file will contain at least enough data to fill the two buffers, but if you plan on using very short files containing less than a few kbytes of data, you might want to modify this code. Check the number of bytes returned by `freadM()` and add the correct padding at the end of the buffer(s).

At this point we are ready to initialize the audio playback "machine," which will be simply our T3Interrupt() function modified to accommodate two channels for stereo playback. We will also add the ability to skip samples, to reduce the sample rate if necessary, and the ability to deal with 16-bit samples (signed) as well as 8-bit samples (unsigned). All this information will be passed to the audio module initAudio() routine as a short list of parameters:

```
// 9. start playing, enable the interrupt
initAudio( wav.srate, skip, size, stereo, fix, pos);
```

As the timer interrupt is activated, the service routine starts immediately consuming data from the first buffer and, as soon as its whole content is exhausted, it will set the flag AEmptyFlag to let us know that new data needs to be retrieved from the WAV file and the second buffer will be selected as the active one. Therefore, to maintain the playback flowing smoothly, we will sit in a tight loop, constantly checking for the AEmptyFlag, ready to perform the refill, counting the bytes we read from the file until we use them all up.

```
// 10. keep feeding the buffers in the playing loop
while (lc >=B_SIZE)
{
    if ( AEmptyFlag)
    {
        r = fread( ABuffer[1-CurBuf], B_SIZE, f);
        AEmptyFlag = FALSE;
        lc-= B_SIZE;

    }
} // while wav data available
```

Actually, we stop a little sooner, when the data left in the file is no longer sufficient to fill another entire buffer load. In that case, unless the data block size was an exact multiple of the buffer size and there is no new data to read, the last piece is loaded and needs to be padded to fill completely what will be the last buffer to play back:

```
// 11. flush the buffers with the data tail
if( lc>0)
{
    // load the last sector
    r = fread( ABuffer[1-CurBuf], lc, f);
    last = ABuffer[1-CurBuf][r-1];
    while(( r<B_SIZE) && (last>0))
        ABuffer[1-CurBuf][r++] = last;

    // wait for current buffer to be emptied
    AEmptyFlag = 0;
    while (!AEmptyFlag);
}
```

We wait then for the completion of the playback of the very last buffer and we immediately terminate the audio playback:

```
// 12. finish the last buffer
AEmptyFlag = 0;
while (!AEmptyFlag);

// 13.stop playback
haltAudio();
```

Closing the file, we release the memory allocated and we return to the calling application:

```
// 14. close the file
fclose( f);

// 15. return with success
return TRUE;
```

```
} // play
```

To complete this module, we need to create a small include file "wave.h" to publish the prototype of the play() function:

```
/*---------------------------------------------------------------------
** Wave.H
**
**  Wave File Player
**  Uses 2 x 8 bit PWM channels
**
*/


unsigned play( const char *name);
```

The low level audio routines

The `play()` function we have just completed relied heavily on a lower-level audio module to perform the actual Timer and OC peripherals initialization, as well as to perform the actual periodic update of the PWM duty cycle. We will call this module "`audiopwm.c`" and it will be mostly based on the code developed in the beginning of this chapter, extended to manage two channels for stereo playback and a number of additional options. The OC1 and OC2 modules will be used simultaneously to produce the left and right channels. The timer interrupt routine will be the real core of the playback functionality. A pointer `BPtr` will keep track of our position inside each buffer, as we will be using up the data to feed the PWM modules with new samples at every period.

```
void _ISRFAST _T3Interrupt( void)
{
    // 1. load the new samples for the next cycle
    OC1RS = 30+(*BPtr ^ Fix);
    if ( Stereo)
        OC2RS =30 + (*(BPtr + Size) ^ Fix);
    else     // mono
        OC2RS = OC1RS;
```

The pointer is advanced by a number of bytes that depends both on the size of the samples (16 or 8 bits each) as well as the need to skip samples to reduce the sample rate when the `play()` function determines it is necessary:

```
    // 2. skip samples to reduce the bitrate
    BPtr += Skip;
```

As soon as a buffer-load of data is used up, we need to swap the active buffer:

```
    // 3. check if buffer emptied
    if ( --BCount == 0)
    {
        // 3.1 swap buffers
        CurBuf = 1- CurBuf;
        BPtr = ABuffer[ CurBuf];
```

Reset the samples counter and set a flag to alert the `play()` routine we need a new buffer to be prepared before we run out of data again:

```
        // 3.2 restart counter
        BCount = B_SIZE/Size;

        // 3.3 flag a new buffer needs to be filled
        AEmptyFlag = 1;
    }
```

Only then can we exit after clearing the interrupt flag:

```
        // 4. clear interrupt flag and exit
        _T3IF = 0;

    } // T3 Interrupt
```

The initialization routine is equally straightforward if you recall the one we created at the beginning of the chapter, except more parameters are passed from the calling application and copied into the module's own (private) variables:

```
void initAudio( long bitrate, int skip, int size, int stereo, int fix, int pos)
{
    // 1. init pointers
    CurBuf = 0;                      // start with buffer0 active first
    BPtr = ABuffer[ CurBuf]+pos;
    BCount = (B_SIZE-pos)/size; // number of samples to be played
    AEmptyFlag = 0;
    Skip = skip;
    Fix = fix;
    Stereo = stereo;
    Size = size;
```

One buffer is selected as the "current" in-use buffer and all the pointers and counters are initialized. Then the timer is initialized and its interrupt mechanism:

```
    // 2. init the timebase
    T3CON = 0x8000;          // enable TMR3, prescale 1:1, internal clock
    PR3 = FCY / bitrate;     // set the period for the given bitrate
    Offset = PR3/2;
    _T3IF = 0;               // clear interrupt flag
    _T3IE = 1;               // enable TMR3 interrupt
```

The duty cycles are initialized next to an initial offset that will be half the value of the period, to provide an even 50% initial output level.

```
    // 3. set the initial duty cycles
    OC1R = OC1RS = Offset;  // left
    OC2R = OC2RS = Offset;  // right
```

Finally, the Output Compare modules are fired up:

```
    // 4. activate the PWM modules
    OC1CON = 0x000E;         // CH1 and CH2 in PWM mode, TMR3 based
    OC2CON = 0x000E;

} // initAudio
```

The function `haltAudio()` called at the end of the playback will definitely be the simplest. Its only task is to disable Timer3 and therefore freeze the Output Compare modules and with them the entire interrupt mechanism:

```
void haltAudio( void)
{
    T3IE = 0;                 // disable TMR3 interrupt
} // halt audio
```

To complete the module you will need the usual header, include files and the definitions of the global variables allocated, which will include the audio buffers.

```
/*
** Audio PWM demo
**
*/


#include <p24fj128ga010.h>

#include "AudioPWM.h"

#define _FAR __attribute__(( far))

// global definitions
unsigned Offset;                    // 50% duty cycle value
char   _FAR  ABuffer[ 2][ B_SIZE];  // double data buffer
int    CurBuf;                      // index of buffer in use
volatile int AEmptyFlag;            // flag a buffer needs to be filled

// internal variables
int Stereo;                         // flag stereo play back
int Fix;                            // sign fix for 16-bit samples
int Skip;                           // skip factor to reduce sample/rate
int Size;                           // sample size (8 or 16-bit)

// local definitions
unsigned char *BPtr;                // pointer inside active buffer
int BCount;
```

Notice that, just as we did in previous lessons, when allocating large buffers for video applications we can use the `far` attribute to allocate memory beyond the PIC24 near addressing space.

349

The include file `"audiopwm.h"` will publish all the necessary definitions and prototypes for the `"Wave.c"` module and other applications to make use of the services provided by the Audio PWM module.

```
/*
** AudioPWM.h
**
*/

#define FCY      16000000L          // instruction cycle frequency
#define TCYxUS   16                 // number of Tcycles in a microsecond
#define B_SIZE   2048               // audio buffer size


extern char ABuffer[ 2][ B_SIZE];   // double data buffer
extern int CurBuf;                  // index of buffer in use
extern volatile int AEmptyFlag;     // flag a buffer needs to be filled

void initAudio( long bitrate, int skip, int size, int stereo, int fix, int pos);
void haltAudio( void);
```

Testing the WAVE file player

Now that the low-level audio module and the playback module have been completed, it is time to put it all together and start testing by playing some music.

Let's create a new project called "WaveTest" and let's immediately add all the necessary modules and their include files to the project. They are:

- `"sdmmc.c"`
- `"fileio.c"`
- `"audiopwm.c"`
- `"wave.c"`
- `"sdmmc.h"`
- `"fileio.h"`
- `"audiopwm.h"`
- `"wave.h"`

Then, let's add a new main module "wavetest.c", which will contain just a few lines of code. It will invoke the play() function indicating the name of a single file that we will have copied onto the SD/MMC card (TRACK00.WAV).

```c
/*
**    WaveTest
**
*/

#include <p24fj128ga010.h>

#include "SDMMC.h"
#include "fileio.h"

#include "../Audio/Audio PWM.h"
#include "../Wave/Wave.h"


main( void)
{

    TRISA = 0xff00;

    if ( !mount())
        PORTA = FError + 0x80;
    else
    {
        if ( play( "TRACK00.WAV"))
            PORTA = 0;
      else
            PORTA = 0xFF;
    } // mounted

    while( 1)
    {
    } // main loop

} //main
```

The PORTA row of LEDs will serve as our display to report errors, should the mount() operation fail or should the file not be found on the storage device.

Build the project and program the code on the Explorer16 board using the appropriate checklists. Don't forget to reserve some room for the Heap, as the fileio.c module uses it to allocate buffers and data structures.

To proceed gradually, I would recommend that you test the program with WAV files of increasingly high sample rates and sizes. For example, you should run the first test with a WAV file using 8-bit samples, mono, at 8k samples/second. Then proceed, gradually increasing the complexity of the format and the speed of playback, possibly aiming to reach with a last test the full capabilities of our application with a 16-bit per sample, stereo, 44,100 samples/second file. The reason for this gradual increase is that we will need to verify that the performance of our "fileio.c" module is up to the task. In fact, as the sample rate, number of channels and size of the samples increase, so does the bandwidth required from the file system. We can quickly calculate the performance levels required by a few combinations of these parameters.

File	Sample size	Channels	Sample-rate	Byte-rate	Reload period (ms)
Voice mono	1	1	8,000	8,000	64.0
Voice stereo	1	2	8,000	16,000	32.0
Audio 8-bit mono	1	1	22,050	22,050	23.2
Audio 8-bit stereo	1	2	22,050	44,100	11.6
Audio 8-bit high bitrate mono	1	1	44,100	44,100	11.6
Audio 8-bit high bitrate stereo	1	2	44,100	88,200	5.8
Audio 16-bit mono	2	1	44,100	88,200	5.8
Audio 16-bit stereo	2	2	44,100	176,400	2.9

The table shows the byte-rate required by each file format—that is, the number of bytes that get consumed by the playback function for every second (sample size × channels × sample rate). In particular, the last column shows how often a new buffer full of data will be required to be replenished (512/byte-rate); that gives us the time available for the play() routine to read the next sector of data from the WAV file.

Notice that since the PIC24 PWMs can only produce less than 9 bits of resolution when operating at the 44,100-Hz frequency, the audio PWM module has been designed to use only the MSB of a 16-bit sample. Therefore, don't expect any increase in the quality of the audio output once you attempt to play back a WAV file in one of the last two formats. All you obtain at that point is a waste of the space on the SD/MMC memory card. If you want to maximize the use of the storage space available, make sure that when copying a file onto the card, you reduce the sample size to 8 bits. You will be able to pack a double number of music files on the card for the same output audio resolution.

If you start experimenting gradually, as I suggested, moving down the table, you should find that beyond a certain point (beyond the audio 8-bit high-bitrate mono, probably) things just won't play out right. The playback will skip, repeat and hiccup and it just won't sound right. What is happening is that the freadM() function has reached its limit and is not capable of keeping up with the audio playback demands. The time it takes, on average, to load a new buffer of data is longer than the time it takes to consume one; after a short while, the `play()` routine starts falling behind and the audio playback function starts repeating a buffer or playing back buffers that are not completely filled yet.

Optimizing the file I/O

When we wrote the file I/O library, and even before, when we wrote the low-level functions to access the SD/MMC card, we have focused mainly on getting things done. We have never really tried to assess the level of performance provided. Perhaps now we have the right motivation to look into it. Throughout the rest of the book we have resisted using any of the optimization features of the compiler, so that every example could be tested using simply the free MPLAB® C compiler Student Version. We want to maintain this commitment. Perhaps there is some room to improve the performance using just a little ingenuity.

The first thing to do is to discover where the PIC24 is spending the most time when reading data from the card. Inspecting the `freadM()` function, you will notice that there are only two calls to lower-level subroutines. One is a `readDATA()` function call used to load a new sector from the current cluster and the other is a `nextFAT()` function call used to identify the next cluster, once every sector of the current cluster is exhausted. Eventually both functions will call in their turn the `readSECTOR()` function to actually retrieve a block of 512 bytes of data. Lastly, a call to the standard C function `memcpy()` is performed to transfer a data block to the calling application buffer. So the ultimate performance of `freadM()` will depend on the performance of `readSECTOR()` and `memcpy()`.

LED Profiling

To determine which one of the two has the largest responsibility is a relatively easy job, if you happen to have an oscilloscope at hand. In fact, if you remember, we designed `readSECTOR()` to use one of the LEDs on PORTA to signal when a read operation is being performed on the SD/MMC card. If we point the oscilloscope on the anode of the corresponding LED during the playback loop, we should be able to see a periodic pulse whose length is indicating to us the exact amount of time the PIC24 is spending inside the `readSECTOR()` function while transferring data. The pause in between the pulses is otherwise proportional to the time spent inside the `memcpy()` function and eventually most of the rest of the `freadM()` function call stack. In one single glance, you will immediately realize where the problem lies.

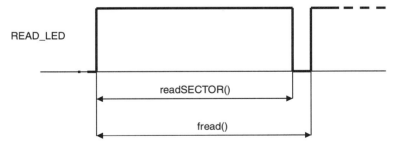

Figure 15-10. Pointing the oscilloscope on the READ_LED pin.

There is no doubt that, it is the `readSECTOR()` function that needs our full attention, since it uses up the largest part of a period that is more than 10 ms long!

```
int readSECTOR( LBA a, char *p)
// a        LBA of sector requested
// p        pointer to sector buffer
// returns  TRUE if successful
{
    int r, tout;

    READ_LED = 1;

    r = sendSDCmd( READ_SINGLE, ( a << 9));
    if ( r == 0)     // check if command was accepted
    {
        // wait for a response
        tout = 10000;
        do{
            r = readSPI();
            if ( r == DATA_START) break;
        }while( --tout>0);

        // if it did not timeout, read a 512 byte sector of data
        if ( tout)
        {
            for( i=0; i<512; i++)
                *p++ = readSPI();

            // ignore CRC
            readSPI();
            readSPI();

        } // data arrived

    } // command accepted

    // remember to disable the card
    disableSD();
    READ_LED = 0;

    return ( r == DATA_START);       // return TRUE if successful
} // readSECTOR
```

If you look at the function listing now, you will notice that there are only three possible areas where the PIC24 could be spending so much time:

1. The `sendSDCmd()` function.

2. The loop where we wait for the DATA_START token from the card (perhaps it is just a slow SD/MMC card?).

3. The loop where we read, one by one, all 512 bytes from the card.

To discriminate among the three we can simply change the point where we turn on the READ_LED and where we turn it off so as to bracket one of three spots. If you recompile the project and run the test a couple of times, you will notice that when bracketing the sendSDCmd() function the pulse on-time is reduced to a barely readable blip.

```
READ_LED = 1;
r = sendSDCmd( READ_SINGLE, ( a << 9));
READ_LED = 0;
```

This means the card is very fast to respond to the command and time must be spent elsewhere.

If you bracket the loop waiting for the DATA_START token, you will get a very similar result:

```
READ_LED = 1;
    // wait for a response
    tout = 10000;
    do{
        r = readSPI();
        if ( r == DATA_START) break;
    }while( --tout>0);
READ_LED = 0;
```

It is the third loop, apparently so innocuous but repeated 512 times, that seems to be taking all the cycles the PIC24 has to spare.

```
READ_LED = 1;
        for( i=0; i<512; i++)
            *p++ = readSPI();
READ_LED = 0;
```

Here is where we will have to concentrate all our optimization efforts.

The first idea that comes to mind is to try to remove the function call to the readSPI() function and replace it directly with the few lines of code required inline:

```
READ_LED = 1;
    for( i=0; i<512; i++)
    {
        SPI2BUF = 0xFF;                       // write to buffer for TX
        while( !(SPI2STATbits.SPIRBF));       // wait for transfer complete
        *p++ = SPI2BUF;                       // read the received value
    }
READ_LED = 0;
```

If you patiently rebuild the project and measure the new pulse length, you should already see an improvement, but it is not going to make enough of a difference.

Looking under the hood for more

The next natural step for us is to take a look at how the compiler is treating those few lines of code, peeking at the specific segment in the disassembly listing window:

```
139:                              for( i=0; i<512; i++)
  011A4  EB0000    clr.w 0x0000
  011A6  980750    mov.w 0x0000,[0x001c+10]
  011A8  9000DE    mov.w [0x001c+10],0x0002
  011AA  201FF0    mov.w #0x1ff,0x0000
  011AC  508F80    sub.w 0x0002,0x0000,[0x001e]
  011AE  3C0013    bra gts, 0x0011d6
  011CE  90005E    mov.w [0x001c+10],0x0000
  011D0  E80000    inc.w 0x0000,0x0000
  011D2  980750    mov.w 0x0000,[0x001c+10]
  011D4  37FFE9    bra 0x0011a8
142:                              {
144:                                  SPI2BUF = 0xFF;
  011B0  200FF0    mov.w #0xff,0x0000
  011B2  881340    mov.w 0x0000,0x0268
146:                                  while( !SPI2STATbits.SPIRBF);
  011B4  BFC260    mov.b 0x0260,0x0000
  011B6  FB8000    ze.b 0x0000,0x0000
  011B8  600061    and.w 0x0000,#1,0x0000
  011BA  E00000    cp0.w 0x0000
  011BC  32FFFB    bra z, 0x0011b4
147:                                  *p++ = SPI2BUF;
  011BE  4701E4    add.w 0x001c,#4,0x0006
  011C0  780093    mov.w [0x0006],0x0002
  011C2  801340    mov.w 0x0268,0x0000
  011C4  784100    mov.b 0x0000,0x0004
  011C6  780001    mov.w 0x0002,0x0000
  011C8  784802    mov.b 0x0004,[0x0000]
  011CA  E80081    inc.w 0x0002,0x0002
  011CC  780981    mov.w 0x0002,[0x0006]
148:                                  }
... <<for loop closing code here>>
  011D6
```

More than 25 instructions are used to perform what seemed a straightforward `for` loop. Naturally, we should look at reducing the complexity of the innermost loop, the `while` loop where we wait for the SPI peripheral to complete the transfer. While most of that loop seems straightforward, there is a sign-extension (`ze.b`) instruction inside it that might appear redundant. It makes us wonder if it is not just a byproduct of bit field arithmetic used by the compiler to check the `SPI2STATbits.SPIRBF` flag.

Reformulating the code using direct masking of the contents of the SPI2STAT register improves the situation and confirms the suspicion:

```
for( i=0; i<512; i++)
{
    SPI2BUF = 0xFF;                  // write to buffer for TX
    while( !(SPI2STAT & 1));         // wait for transfer complete
    *p++ = SPI2BUF;                  // read the received value
}
```

The code produced is now just one instruction shorter, but keep in mind that that instruction is repeated at least twice for each of the 512 loops.

```
139:                        for( i=0; i<512; i++)
 011A4  EB0000    clr.w 0x0000
 011A6  980750    mov.w 0x0000,[0x001c+10]
 011A8  9000DE    mov.w [0x001c+10],0x0002
 011AA  201FF0    mov.w #0x1ff,0x0000
 011AC  508F80    sub.w 0x0002,0x0000,[0x001e]
 011AE  3C0012    bra gts, 0x0011d4
 011CC  90005E    mov.w [0x001c+10],0x0000
 011CE  E80000    inc.w 0x0000,0x0000
 011D0  980750    mov.w 0x0000,[0x001c+10]
 011D2  37FFEA    bra 0x0011a8
142:                          {
144:                              SPI2BUF = 0xFF;
 011B0  200FF0    mov.w #0xff,0x0000
 011B2  881340    mov.w 0x0000,0x0268
145:                              while( !(SPI2STAT&1));
 011B4  801300    mov.w 0x0260,0x0000
 011B6  600061    and.w 0x0000,#1,0x0000
 011B8  E00000    cp0.w 0x0000
 011BA  32FFFC    bra z, 0x0011b4
146:                              *p++ = SPI2BUF;
 011BC  4701E4    add.w 0x001c,#4,0x0006
 011BE  780093    mov.w [0x0006],0x0002
 011C0  801340    mov.w 0x0268,0x0000
 011C2  784100    mov.b 0x0000,0x0004
 011C4  780001    mov.w 0x0002,0x0000
 011C6  784802    mov.b 0x0004,[0x0000]
 011C8  E80081    inc.w 0x0002,0x0002
 011CA  780981    mov.w 0x0002,[0x0006]
147:                          }
... <<for loop closing code here>>
 011D4
```

The next trick up our sleeve is to try to reduce the shuffling of data to and from the software stack by assigning specific registers to hold variables of frequent use. One such candidate is the variable `i` used as an index in the `for` loop, and the other one is the pointer `p`.

The C30 compiler will let us assign a variable to a register with the following syntax:

```
register unsigned i asm( "w5");
```

but the result is not guaranteed unless the specific register is available. Typically the compiler uses the first four registers `W0...W3` as a scratch pad and won't let us have one of those all for ourselves. Also, the register cannot be a parameter of the function, as unfortunately is the case for `p`, as this might impact the register allocation scheme of the calling functions. We can quickly work around such limitations by copying the contents of `p` into a new pointer that we will call `q`, as in the following code:

```
register unsigned i asm( "w5");
register char * q asm("w6");
q = p;
for( i=0; i<512; i++)
{
    SPI2BUF = 0xFF;
    while( !(SPI2STAT&1));  // wait for transfer to complete
    *q++ = SPI2BUF;                // read the received value
}
```

This time, recompiling the code, we can observe a considerable reduction in the outer loop size and a simplification in the `for` loop encoding:

```
139:                              for( i=0; i<512; i++)
  011A6   EB0280    clr.w 0x000a
  011A8   201FF0    mov.w #0x1ff,0x0000
  011AA   528F80    sub.w 0x000a,0x0000,[0x001e]
  011AC   3E000D    bra gtu, 0x0011c8
  011C4   E80285    inc.w 0x000a,0x000a
  011C6   37FFF0    bra 0x0011a8
142:                                  {
144:                                      SPI2BUF = 0xFF;
  011AE   200FF0    mov.w #0xff,0x0000
  011B0   881340    mov.w 0x0000,0x0268
145:                                      while( !(SPI2STAT&1));
  011B2   801300    mov.w 0x0260,0x0000
  011B4   600061    and.w 0x0000,#1,0x0000
  011B6   E00000    cp0.w 0x0000
  011B8   32FFFC    bra z, 0x0011b2
146:                                      *q++ = SPI2BUF;
  011BA   801340    mov.w 0x0268,0x0000
  011BC   784080    mov.b 0x0000,0x0002
  011BE   780006    mov.w 0x000c,0x0000
  011C0   784801    mov.b 0x0002,[0x0000]
  011C2   E80306    inc.w 0x000c,0x000c
...  <<for loop closing code here>>
  011C8
```

We are down to 17 instructions. The last step will consist of trying to use a different type of loop to count the 512 bytes of data. This time we will use a simple `do` loop and we will count backward:

```
register unsigned i asm( "w5");
register char * q asm("w6");
q = p;
i = 512;
do {
    SPI2BUF = 0xFF;
    while( !(SPI2STAT&1)); // wait for transfer to complete
    *q++ = SPI2BUF;              // read the received value
} while ( --i>0);
```

This gives us the best results so far—only 15 instructions all included!

```
  011A6   202005      mov.w #0x200,0x000a
141:                             do{
144:                                 SPI2BUF = 0xFF;
  011A8   200FF0      mov.w #0xff,0x0000
  011AA   881340      mov.w 0x0000,0x0268
145:                                 while( !(SPI2STAT&1));
  011AC   801300      mov.w 0x0260,0x0000
  011AE   600061      and.w 0x0000,#1,0x0000
  011B0   E00000      cp0.w 0x0000
  011B2   32FFFC      bra z, 0x0011ac
146:                                 *q++ = SPI2BUF;
  011B4   801340      mov.w 0x0268,0x0000
  011B6   784080      mov.b 0x0000,0x0002
  011B8   780006      mov.w 0x000c,0x0000
  011BA   784801      mov.b 0x0002,[0x0000]
  011BC   E80306      inc.w 0x000c,0x000c
148:                                 } while (--i>0);
  011BE   E90285      dec.w 0x000a,0x000a
  011C0   E00005      cp0.w 0x000a
  011C2   3AFFF2      bra nz, 0x0011a8
  011C4
```

It is time to reprogram the Explorer16 board with the new code and check once more with the scope to see how long it takes now for the readSECTOR() function to complete reading a 512-kbyte sector of data. You will be pleasantly surprised to verify that we have now managed to reduce the time required to less than 1.5 ms. This will be enough to let us play back even the most demanding WAV file and then some.

Post-flight briefing

This final lesson was perhaps the ideal conclusion for our learning experience, as we mixed the most advanced software and hardware capabilities in a project that covered both the digital and the analog domain. We started using the Output Compare peripherals to produce analog signals in the audio spectrum of frequencies. We used this new capability together with the "fileio.c" module, developed in the previous lesson, to play back uncompressed music files (WAV file format) from a mass-storage device (SD/MMC card). The basic media player application obtained represents only a new starting point. There is no limit to the possible expansions of this project and if I have managed to excite your curiosity and imagination, there is no limit to what you can do with the PIC24 and the MPLAB C30 compiler.

Tips and tricks

The beginning and the end of the playback are two critical moments for the PWM modules. At rest, the output filter capacitor is discharged and the output voltage is 0V. But as soon as the playback begins, a 50% duty cycle will force it to ramp very quickly to approximately a 1.5V level, producing a loud and unpleasant click. The opposite might happen at the end should we turn off the PWM modules instead of just disabling the interrupts as we did in the demo project. The phenomenon is not dissimilar to what happens to analog amplifier circuits at power-on and -off. A simple work-around consists of adding just a couple of lines of code. Before the timer interrupt is enabled and the playback machine starts, add a small (timed) loop to gradually increase the output's duty cycle from zero all the way up to the value of the first sample taken from the playback buffer.

Exercises

1. Investigate the decoding techniques for ADPCM signals for use with voice messages (see application note AN643).

2. Search for all the "WAV" files on the card, and build a "playlist."

3. Implement a "shuffle" mode using the pseudo-random number generator and gradually emptying the playlist.

4. Experiment with basic digital filtering techniques to remove undesired frequencies, boost others or simply distort sounds and voices.

Books

- Mandrioli, D. & Ghezzi, C. (1987)

  Theoretical Foundations of Computer Science

  John Wiley & Sons, New York, NY

  Not an easy read, but if you are curious about the deep mathematical, theoretical foundations of computer science…

- Leroy Cook, (1990)

  101 Things to Do With Your Private License

  TAB books, a division of McGraw-Hill, Inc

Links

- *http://en.wikipedia.org/wiki/RIFF*

  The RIFF file format explained.

- *http://en.wikipedia.org/wiki/WAV*

  The WAVE file format explained.

- *http://ccrma.stanford.edu/courses/422/projects/WaveFormat/*

  Another excellent description of the WAVE file format.

About the Author

Lucio Di Jasio received his MSEE (Summa cum Laude) from the University of Trieste, Italy in 1990, presenting a thesis on the "Simulation of digital logic circuits using the Occam model of parallelism." After graduating, he worked as a software/hardware designer on projects as diverse as Parallel C digital-image processing in industrial-automation applications, Unix C/4GL programming in Supervisory Control And Data Acquisition (SCADA) applications, and encryption for security systems in automotive applications.

He joined Microchip Technology in 1995 as a Field Application Engineer covering the South of Europe. In 2000, he moved to Chandler, AZ, and specialized in the KEELOQ® secure-data product line, publishing several application notes.

In 2002, Lucio moved into a Product Marketing position, supporting the definition and launch of the new High Pin Count and High Density families of PIC microcontrollers. Since 2005, he has been in charge of the Application Segment Group, a cross-divisional team of engineers that develops and promotes Microchip's solutions across a wide range of application segments, including: utility metering, intelligent power conversion, motor control and lighting applications.

Lucio earned his private pilot license in 2002, and an instrument rating in 2005. He has accumulated 350 hours of experience in various single engine airplanes. Lucio owns a Cessna 172 (N75816), which he tries to fly as frequently as possible to escape the heat of the Arizona summer.

Index